Routledge Introductions to Development

Series Editors:
John Bale and David Drakakis-Smith

Women and Development in the Third World

For all societies, the common denominator of gender is female sub-ordination. For women of the contemporary Third World the effects of patriarchal attitudes are exacerbated by economic crisis and the legacy of imperialism.

Feminist critique has introduced the gender factor to development theory, arguing that the equal distribution of the benefits of economic development can only be achieved through a radical restructuring of the process of development. Now, the universal validity of both gender-neutral development theory and the feminist concepts of the post-industrial world are being questioned.

All societies establish a clear-cut division of labour by sex. Moderniz-ation and restructuring of traditional economies have altered this division of labour, often increasing women's dependent status as well as their work-load. Women in the Third World cope with a double or triple burden of productive and reproductive work and everywhere put in longer hours than men. Janet Henshall Momsen presents ten worldwide case studies personalized examples of women's lives and coping strategies in the Third World. Her review of policy and practice raises questions about development planning and the empowerment of women. The book concludes with a discussion of the impact of environmental degradation and economic restructuring on women, describing the integral position of women in any solution to the current crises facing the Third World.

Women and Development comes as a welcome introductory text in an area where the present literature is concentrated on the advanced and the specialized. The book will be invaluable to students of Develop-ment, Geography, Sociology, Economics and Women's Studies and of interest to all concerned for the position of women in the world today.

A volume in the **Routledge Introductions to Development** series edited by John Bale and David Drakakis-Smith.

In the same series

John Cole
Development and Underdevelopment
A Profile of the Third World

David Drakakis-Smith
The Third World City

Allan and Anne Findlay
Population and Development in the Third World

Avijit Gupta
Ecology and Development in the Third World

John Lea
Tourism and Development in the Third World

John Soussan
Primary Resources and Energy in the Third World

Chris Dixon
Rural Development in the Third World

Alan Gilbert
Latin America

David Drakakis-Smith
Pacific Asia

Rajesh Chandra
Industrialization and Development

Mike Parnwell
Population Movements and the Third World

Janet Henshall Momsen

Women and Development in the Third World

London and New York

First published 1991
by Routledge
11 New Fetter Lane, London EC4P 4EE

Simultaneously published in the USA and Canada
by Routledge
29 West 35th Street, New York, NY 10001

Reprinted 1993

Typeset by J&L Composition Ltd, Filey, North Yorkshire.
Printed and bound in Great Britain by
Biddles Ltd, Guildford and King's Lynn

British Library Cataloguing in Publication Data
Momsen, Janet Henshall
Women and development in the Third World – (Routledge
introductions to development) 1. Economic development. Role of
women. Social aspects I. Title
330.91724

ISBN 0–415–01695–9

Library of Congress Cataloging in Publication Data
Momsen, Janet Henshall.
Women and development in the Third World/Janet Henshall Momsen.
p. cm. — (Routledge introductions to development)
Includes bibliographical references and index.
ISBN 0–415–01695–9
1. Women in development—Developing countries—Case studies.
2. Sexual division of labor—Developing countries—Case studies.
I. Title. II. Series.
HQ1240.5.D44M66 1991
307.1′412′091724—dc20   90–43653
                                                  CIP

To my mother who, as she approaches her ninth decade, continues to provide enthusiastic support.

Acknowledgements

This book, more than most, owes much to other people. I should like to thank my students at Newcastle University and my friends and colleagues Christine Barrow, Janice Monk, Liz Oughton and Janet Townsend for discussing and challenging my ideas. I gained much from the contributions of the members of the Institute of British Geographers Short-Life Working Group on Women and Development and from the participants from twenty-nine countries who came to the Commonwealth Geographical Bureau Workshop on Gender and Development held at Newcastle in 1989. Above all I thank the many rural Third World women who have given of their precious time to answer my questions and provide enlightenment during fieldwork over the last twenty-seven years. The mistakes remain my own.

I am grateful to Ann Rooke for her skilful execution of the diagrams and maps.

Finally I should like to thank my sons Richard and Magnus for their support during the birth pangs of the manuscript despite the disruptions it caused in the household routine.

Contents

Plates

Figures

Tables

1
Introduction

The development process affects women and men differentially. The after effects of colonialism and the peripheral position of Third World countries in the world economy exacerbate the effects of sexual discrimination on women. The penetration of capitalism, leading to the modernization and restructuring of traditional economies, often increases the disadvantages suffered by women as the modern sector takes over many of the economic activities, such as food processing and making of clothes, which had long been the means by which women supported themselves and their families. A majority of the new and better-paid jobs go to men but male income is less likely to be spent on the family.

Modernization of agriculture has altered the division of labour between the sexes, increasing women's dependent status as well as their workload. Women often lose control over resources such as land and are generally excluded from access to new technology. Male mobility is higher than female, both between places and between jobs, and more women are being left alone to support children. Women in the Third World now carry a double or even triple burden of work as they cope with housework, childcare and subsistence food production, in addition to an expanding involvement in paid employment. Everywhere women work longer hours than men. How women cope with declining status, heavier work burdens and growing impoverishment is crucial to the success of development policies in the Third World.

Women constitute almost half the world's population but even today there are 80 million fewer girls than boys enrolled in school. Women carry the burden of two-thirds of the total hours of work performed. For this they earn a mere 10 per cent of the world's income and own but 1 per cent of the property. Women produce more than half of the locally-grown food in developing countries and as much as 80 per cent in Africa.

Within these broad generalizations women's lives in different places show great variation: most typists in Martinique are women but this is not so in Madras, just as women make up the vast majority of domestic servants in Lima but not in Lagos. Nearly 90 per cent of sales workers in Accra are women but this proportion falls to a bare 1 per cent in Algeria. In every country, the jobs done predominantly by women are the least well paid and have the lowest status. Clearly female and male roles are neither equal nor fixed. They differ from place to place and this spatial variation is most marked in the Third World. The relationship between these spatial patterns and development is the theme of this book.

Forty years ago, in 1948, the Universal Declaration of Human Rights reaffirmed the belief in the equal rights of men and women, first laid down by the nations of the world in the Charter of the United Nations. Today it is clear that progress towards equality for women in most parts of the world is considerably less than that which was promised. However, disparities between women in different countries are greater than those between men and women in any one country. Life expectancy at birth for women varies from 74 years in Cuba to 43 in Chad. The proportion of illiterates in the female population varies from 99 per cent in Ethiopia to less than 1 per cent in Barbados. Even within individual countries women are not a homogeneous group but can be differentiated by class, ethnicity and life stage. Thus the range on most socio-economic measures is wider for women than for men and is greatest among the countries of the Third World.

We have now reached the end of the United Nations Third Development Decade while the Decade for Women culminated in a conference in Nairobi in 1985. At the conclusion of the first two Development Decades it was found that the extent of poverty, disease, illiteracy and unemployment in the Third World had increased. During the 1980s we have witnessed unprecedented growth of Third World debt and acute famine in Africa. Similarly the Decade for Women saw only very limited changes in patriarchal attitudes, that is institutionalized male dominance, and few areas where modernization was associated with a reversal of the overwhelming subordination of women.

Yet despite the apparent lack of change, the United Nations Decade for Women achieved a new awareness of the need to consider women when planning for development. In the United States the Percy Amendment of 1973 ensured that women had to be specifically included in all projects of the Agency for International Development. The British Commonwealth established a Woman and Development programme in 1980 supported by all member countries. In many parts of the Third World women's organizations and networks at the community and national level have come to play an increasingly important role in the initiation and implementation of development projects. Above all, the Decade for Women brought about a realization that data collection and research were needed in order to document the situation of women throughout the world. The consequent outpouring of information has made this book possible.

Women and development

Prior to 1970 it was thought that the development process affected men and women in the same way. Productivity was equated with the cash economy and so most of women's work was ignored. When it became apparent that economic development did not automatically eradicate poverty through trickle-down effects, the problems of distribution and equality of benefits to the various segments of the population became of major importance in development theory. Research on women in Third World countries challenged the most fundamental assumptions of international development, added a gender dimension to the study of the development process, and demanded a new theoretical approach.

The early 1970s' model of 'integration', based on the belief that women could be brought into existing modes of benevolent development without a major restructuring of the process of development, has been the object of much feminist critique. The alternative vision, recently put forward, of development *with* women, demands not just a bigger piece of someone else's pie, but a whole new dish, prepared, baked and distributed equally. International development has been challenged to transform itself into a process that is both human-centred and environmentally conservationist.

The principal themes

Three fundamental themes have emerged from the literature on women and development. The first is the realization that all societies have established a clear-cut division of labour by sex, although what is considered a male or female task varies cross-culturally, implying that there is no natural and fixed gender division of labour. Secondly, research has shown that in order to comprehend gender roles in production, we also need to understand gender roles within the household. The integration of women's reproductive and productive work within the private sphere of the home and in the public sphere outside must be considered if we are to appreciate the dynamics of women's role in development. The third fundamental finding is that economic development has been shown to have a differential impact on men and women and the impact on women has, with few exceptions, generally been negative. These three themes will be examined in the chapters that follow.

The overall framework of the book is provided by spatial patterns of gender. Gender is a social phenomenon, socially constructed, while sex is biologically determined. Gender may be derived, to a greater or lesser degree, from the interaction of material culture with the biological differences between the sexes. Since gender is created by society its meaning will vary from society to society and will change over time. Yet, for all societies, the common denominator of gender is female subordination, although relations of power between men and women may be experienced and expressed in quite different ways in different places and at different times. Spatial variations in the construction of gender are considered at several scales of analysis, from continental patterns, through national and regional variations, to the interplay of power between men and women at the household level.

Table 1.1 provides a macro-scale view of women's position on various indicators for countries grouped according both to income level and to location in the Third World. Low-income countries are characterized by populations in which women form a majority. These women bear many children and are poorly educated but undertake a high proportion of the work, especially in agriculture. In most cases, as national income increases, the sex ratio becomes more balanced and women have fewer children, are better educated and do less agricultural work. Yet the high-income, oil-exporting countries of the Middle East and North Africa have predominantly male populations because of the immigration of male workers, while the women have high fertility rates and low levels of economic activity despite relatively high participation in tertiary education.

On a continental scale Latin America is distinguished by high levels of

Table 1.1 Regional patterns of gender differences

World Bank group	Sex ratio	Female life expectancy	Total fertility rate	Female literacy rate	Female/ male adult literacy rate	% of females in tertiary education	Female participation rate in the labour force	Female share of the agricultural labour force
Low-income	98.7	53.6	5.8	37.7	39.5	3.0	47.5	45.5
Low middle-income	100.4	63.8	4.9	54.3	59.5	8.1	37.8	32.0
Upper middle-income	101.4	66.9	3.8	71.1	72.6	13.8	38.2	28.9
High-income oil exporters	145.7	65.2	5.5	52.8	68.4	13.2	28.5	26.0
World region								
Latin America	98.6	68.8	4.0	81.9	77.6	16.6	24.8	17.6
Caribbean	97.0	66.8	3.9	75.0	65.0	8.5	49.5	41.8
Middle East & North Africa	104.5	62.6	5.5	44.7	58.6	8.5	33.9	30.7
Asia	102.9	59.9	4.3	47.8	52.5	5.4	40.9	45.0
Sub-Saharan Africa	97.8	53.1	6.2	37.3	36.1	2.1	47.3	46.4

Sources: The World Bank, *World Development Report, 1985* New York: Oxford University Press. R. L. Sivard, (1985) *Women . . . a world survey,* Washington, D.C.: World Priorities. J. Seager and A. Olson, (1986) *Women in the World,* London: Pan Books

female literacy but low levels of participation by women in the formal workforce, especially in agriculture. This is almost a mirror image of the situation in Africa, south of the Sahara, where women play a major role in agricultural production but suffer from low levels of literacy and life expectancy. The interrelationships between these indicators will be examined in the following chapters.

Key ideas

1 All social groups have developed a division of labour by sex but this varies cross-culturally.
2 Economic development has tended to make the lives of the majority of women in the Third World more difficult.
3 The universal validity of both gender-neutral development theory and of feminist concepts derived from white, Western middle-class women's experience is being questioned.
4 Indicators of quality of life show great variation between countries and between women and men.
5 Measures describing the role and status of women display distinct regional patterns.

2
The sex ratio

It might be expected that the sex ratio, or the proportion of women and men in the population, would be roughly equal everywhere. Figure 2.1 shows that this is not so and there is quite marked variation between countries. Explanations of these spatial patterns reveal differences both in the relative status accorded to women and men and in the quality of life they enjoy in the Third World.

More males than females are conceived but women tend to live longer than men for hormonal reasons. Boys are more vulnerable than girls both before and after birth. The better the conditions during gestation, the more boys are likely to survive and the sex ratio at birth is usually masculine. However, if basic nutrition and health care is available to the whole population, age-specific death rates favour women. In the industrial market economies these factors have resulted in ratios of about 95 to 97 males per 100 females in the general population. Sex-specific migration or warfare may distort the normal demographic pattern. Typically, however, in the absence of such factors, a male-female ratio significantly above 100 reflects the effects of discrimination against women.

In the world as a whole there are some 20 million more men than women because of masculine sex ratios in the Middle East and North Africa, and the very marked imbalance in the huge populations of China and India, where there are 21 million more Chinese men and 27 million more Indian men than women. Between 1965 and 1987 the sex ratio in some countries such as Canada, Australia, the United States, Kuwait,

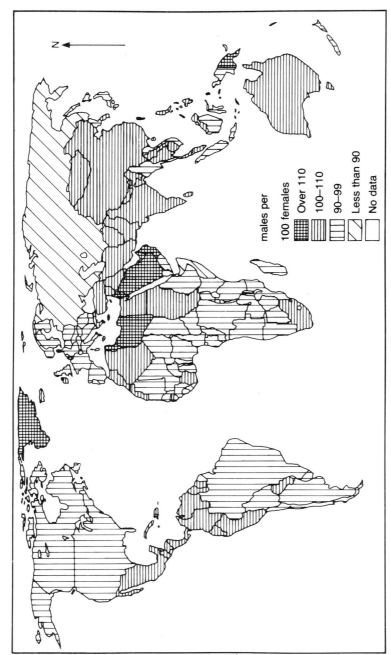

Figure 2.1 Sex ratios

Somalia and Sri Lanka became more feminine but in the world as a whole it became more masculine. In this chapter we examine the reasons for these differences.

Survival

Life expectancy at birth is the most useful single indicator of female general well-being in the Third World. In the developed world average female life expectancy at birth varies only between 81 years in Japan and 73 in Romania, but in the developing world the range extends from 79 years for the women of Hong Kong to 37 in Afghanistan. Women have the shortest lives in the countries of tropical Africa and South Asia. Countries such as China, Haiti and Somalia, with similar per capita Gross National Products to that of Afghanistan of approximately US $300 per year, have female life expectancies of 71, 56 and 49 years respectively. These figures demonstrate that even poor countries can improve the general well-being of their women citizens by adopting a basic needs approach and ensuring that food, health care and education are accessible to all. Within countries marked regional differences may exist: female life expectancy in Malaysia was 59 years in 1965 rising to 72 in 1987 while male life expectancy increased from only 56 to 68 years, but in the east-Malaysian state of Sabah female life expectancy in 1970 at a mere 45 years was three years less than that of men.

Male and female survival chances vary at different points in their life cycle. In the first year of life boys are more vulnerable than girls to diseases of infancy and in old age women tend to live longer as they are less likely to suffer from heart disease. Any deviations from these norms indicate location and culture-specific factors. This can be illustrated by reference to sex ratios at different ages for Libya, a relatively rich country with an economy based on the export of petroleum. Figure 2.2 shows the population pyramid for Libyan citizens in 1973. At every age there is a masculine sex ratio. Poor maternity care is revealed in higher death rates for women in the early and late years of childbearing when risk to the mother is greatest. This contributes to the unusual pattern of an increase in the proportion of men in the population with age. This increase is also explained by under-reporting of the female population and by the repatriation of Libyan males attracted back from overseas by the booming Libyan economy.

Table 2.1, showing age-specific sex ratios during a period of very rapid economic and political change in Libya, illustrates the dynamic

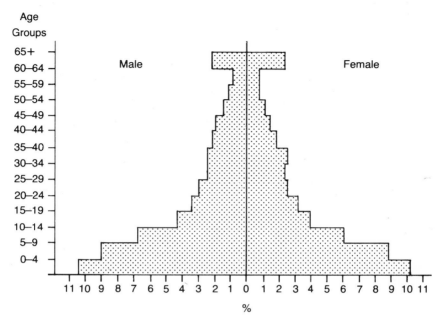

Figure 2.2 Libya: age and sex structure, 1973

nature of sex ratios. The data for 1954 reveal a traditional pattern of high child mortality and low life expectancy, both of which females survive better than males. High maternal mortality and short-term male migration interact to produce very different sex ratios at different stages in the population life cycle. With development, the infant mortality rate fell from about 300 per 1000 births in 1954 to 39 in 1978 as oil revenues came to dominate the national economy. The consequent improved survival level of male babies is reflected in the child sex ratio from 1964 onwards. The decline in the masculine bias of the sex ratio for all ages by 1985 is the result of improvements in statistical reporting and maternal health care, both of which are related to Colonel Qadhafi's efforts to raise the status of women in a highly patriarchal, traditional society and to a better distribution of the benefits of oil riches.

The sex ratio in South Asia

In the early 1980s only six countries in the world recorded a lower life expectancy for women than for men (Figure 2.3). Five of these countries

Table 2.1 Libya: male–female sex ratios 1954–85

Age group	1954	1964	1973	1985*
0–4	99.8	102.9	102.8	102.6
5–9	102./	104.3	102.5	102.3
10–14	120.3	116.4	109.5	103.1
15–19	112.4	104.0	110.3	103.9
20–24	114.7	109.4	107.7	105.4
25–29	107.9	101.3	100.2	106.6
30–34	103.2	106.3	108.9	107.6
35–39	109.2	116.4	101.2	108.0
40–44	89.7	105.8	115.3	108.7
45–49	118.1	111.2	114.0	108.1
50–54	104.9	113.7	112.6	107.4
55–59	157.8	123.7	119.3	106.0
60–64	104.4	127.6	104.2	103.0
65–69	140.0	119.7	119.5	100.0
70+	99.5	128.5	107.3	92.9
All ages	107.8	108.5	106.4	104.2

* Because the 1984 figures were not available, estimates for 1985 were used instead.
Source: A. O. Ibrahim (1987) *The Labour Force in Libya: Problems and Prospects*, unpublished Ph.D. thesis, University of Durham, England

– Bangladesh, Bhutan, India, Nepal and Pakistan – are in South Asia. The sixth country was Papua New Guinea where male life expectancy, estimated to be 50.5 years, was thought to be six months longer than female. Indications are that this discrepancy has now disappeared. The greatest difference is found in Pakistan where men, on average, may expect to live two years longer than women.

In South Asia masculine sex ratios have become more extreme over time with the ratio for India increasing from 103 males per 100 females in 1901 to 107 in 1981. Spatial contrasts are very marked and have remained stable for a long period. With the exception of the small populations of the hill states, the sex ratio is most masculine in the north and the west of the region while the south and east have more balanced or feminine ratios. Masculine sex ratios are associated with high mortality rates for young girls and for women during the childbearing years. It has been calculated that if the African sex ratio existed in India there would have been nearly 30 million more women in India than actually live today. This situation in South Asia has been linked to the general economic undervaluation and low social status of women in the region.

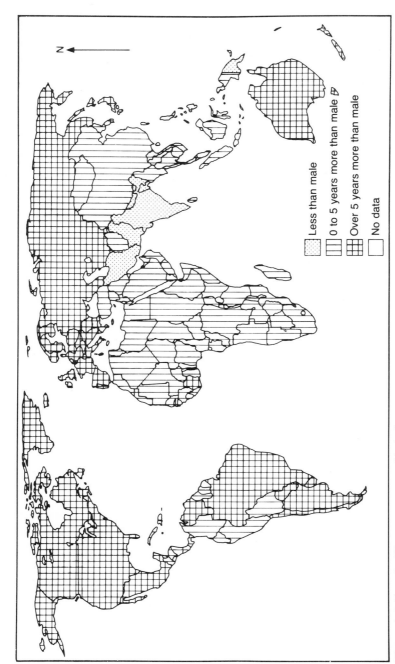

Figure 2.3 Female life expectancy at birth, in years

Legend:
- Less than male
- 0 to 5 years more than male
- Over 5 years more than male
- No data

Nutrition

Low status is reflected in poor female nutrition levels which make women more vulnerable to disease. Men and boys are fed first and receive the major allocation of whatever protein is available (see Case study A). It has been shown that in rural Indian families the percentage contribution to the human energy needs of the household by women, men and children is 53, 31 and 16 per cent respectively. Among the poorest castes as many as half the females but only 14 per cent of males have body weights which are 70 per cent or less of expected weight.

If they fall ill men are more likely than women to receive medical assistance. Illness in young girls and women is often fatalistically accepted by family members. Female infanticide has long been a tradition in many states in northern India. Indeed it has been suggested that in some poor families, mothers feel that their daughters are better off dying as children than growing up to suffer as they themselves have. Overworked, undernourished and anaemic women tend to produce smaller babies and to be more vulnerable to the dangers of childbirth. Maternal death rates are exacerbated by the dominance of traditional medicine in obstetrics and gynaecology in many parts of the region. Death rates for girls between the ages of 1 and 4 in north India and Pakistan have in some cases been found to be almost twice as high as for boys in the same age group while in south India and Bangladesh these differences, although still significant, are less. Famine tends to exacerbate these differential mortality rates especially for the poor.

Case study A

Bangladesh – discrimination within a slum household

In 1984, Mina, a Hindu widow aged about 40, was living with her four children aged 12, 10, 7, and 27 months, in a slum in the town of Khulna in Bangladesh. She had had a gastric ulcer for the last ten years but could not afford medical treatment. Her Body Mass Index (BMI), based on her weight and height, was 12.4. The BMI measures leanness in adults and a level of less than 20 is thought to indicate severe malnutrition and a high risk of mortality. Her son aged 12 had a weight for height measurement of 96 per cent of the expected level. Her three daughters had levels of between 64 per

Case study A *(continued)*

cent and 75 per cent of the expected weight for height. Levels below 80 per cent indicate second and third degree malnutrition. Early in 1985 the three girls were admitted to a nutrition rehabilitation centre where they were diagnosed as severely malnourished with upper respiratory diseases, worms and deficiency diseases associated with lack of iron and vitamins A, B and C.

Mina worked 91 hours a week as a domestic servant while her son worked 35 hours a week as a bread seller. Not only was he the only member of the household not severely undernourished but he was also the only one of the children to attend the local free community school. The total income for the household was £11 a month which was about one-quarter of the average income for a slum household of the same size.

Mina and her children reached this state because of the patriarchal social system which tied her economic fortunes first to those of her father and later to those of her husband and her employer. Her father was disinherited because of family quarrels and so without access to land he was forced, after a period of mental illness, to migrate to the city when Mina was two years old. From a very young age she helped to support the family by working as a servant. When she was 12 a marriage was arranged to a distant cousin fifteen years older than her. He owned land and was skilled as a potter but in the Liberation War, as Hindus, they lost everything and had to flee to a refugee camp in India. In 1972, on their return to Bangladesh, they found themselves without a means of livelihood and were forced to move in with her parents in the slum in Khulna. By 1979 it was clear that her husband had contracted tuberculosis. He refused medical help on religious grounds and would not allow Mina to seek health care for their children. Two of her children, a girl and a boy, died within a month of their birth. In 1981 her father was killed in a road accident and in 1984 her husband died. Under pressure from relatives he had sought help from spiritual healers, homeopaths and allopaths during the last few months of his life at a cost of £86.

During the period of her husband's illness and afterwards, Mina was entirely dependent on her patron who was both her landlord and her employer. He paid her less than 50 per cent of the market

Case study A *(continued)*

> rate for domestic service but provided the money for her to
> have hospital treatment when she was suffering from severe
> dysentery, and occasionally gave the children food. Without his
> help the family might not have survived and so she was deferential
> to him and accepted his paternalism. Mina protected her son from
> the worst effects of the struggle for survival as she knew that only
> through his success could the family hope for an improvement.
>
> *Source:* J. Pryer (1987) 'Production and reproduction of malnu-
> trition in an urban slum in Khulna, Bangladesh' in J. H. Momsen
> and J. Townsend (eds.) *Geography of Gender in the Third World*,
> London: Hutchinson, pp. 117–31

Economic status

Urban employment opportunities for women in industry, trade and
commerce are contracting and in rural areas technological change is
reducing their role in agriculture, especially in the processing of crops.
This decline in the economic role of women can be linked to increased
discrimination against them. However, the relationship between
women's role in production and the sex ratio is neither simple nor
universal.

Another explanation of regional differences in the sex ratio of the
Indian population is based on north-south contrasts in the transfer of
property on marriage and at death. In the north, where the sex ratio is
most masculine, not only are women excluded from holding property
but they also require dowries on marriage and so are costly liabilities.
This has led a growing number of pregnant women from all classes
mainly in north-west India to utilize modern medical technology, developed to detect genetic deformities in foetuses, to determine the sex of
the unborn child so that female foetuses can be aborted. Between 1978
and 1983 78,000 female foetuses were aborted because the cost of a sex
determination test followed by abortion of the 'dispensable sex' was less
than the cost of a dowry. Techniques for sex-preselection are also
becoming popular in order to reduce the number of females conceived.
Dowry deaths are increasing in the face of the rising expectations of
husbands' families who now demand such items as television sets and

refrigerators as well as the more traditional jewellery and land. When the new daughter-in-law's dowry fails to fulfil expectations she may find herself burnt to death in a kitchen 'accident'. Sons, on the other hand, contribute to agricultural production, carry the family name and property, attract dowries into the household and take care of parents in their old age.

In the south women may inherit property and their parents may sometimes demand a brideprice from the husband's family, although dowries are becoming more common than in the past. Generally, in southern India women play a greater economic role in the family, the sex ratio is more balanced, fewer small girls die and female social status is higher than in the north. The position of women is most favourable in the south-western state of Kerala where a traditional matriarchal society allowed women greater autonomy in marriage and a long history of activity by Christian missionaries has helped to ensure that women are less discriminated against in access to education than elsewhere in India. The women of Kerala, with the help of women doctors, took family planning into their own hands and very quickly reduced the birthrate without government interference.

Regional patterns of sex ratios in South Asia are highly complex and vary with caste and culture. Most women have little autonomy or access to power or authority. They are faced by discrimination and exclusion and also by oppressive practices such as widow burning, known as suttee, which appears to be on the increase. These social constraints owe their origin to the need to protect the family lineage through the male line by controlling the supply of women. Their effect is most severe at those times in a woman's life when she is particularly physiologically vulnerable, that is below the age of five and during the childbearing years.

Sri Lanka

However, it should be noted that in one country in South Asia women do normally live longer than men. Sri Lanka's development process has included far-reaching social welfare programmes, especially free education and health care, for the last four decades and the benefits can clearly be seen in the improvement in life expectancy (Table 2.2). By 1967 female life expectancy, which had been two years less than that of men twenty years earlier, had surpassed male life expectancy by two years. Twenty years on Sri Lankans of both sexes have the highest life expectancy in South Asia and the additional years women may expect to live have in the last decade suddenly increased from two to five, although male life expectancy has fallen reflecting high male mortality levels in the recent civil disturbances. Yet Sri Lanka still has a masculine sex ratio, despite the improvement in female mortality rates.

Table 2.2 Sri Lanka: expectation of life at birth in years

	1920–2	1946	1953	1963	1967	1981	1987
Male	32.7	43.9	58.8	61.9	64.8	69.2	68.0
Female	30.7	41.6	57.5	61.4	66.9	71.7	73.0

Source: Department of Census and Statistics, Sri Lanka, for the period 1920 to 1981, and World Bank, *Development Report 1989* for 1987 data

Migration

Sex specific migration also affects sex ratios. In Libya, for the last two decades, a booming economy suffering from a labour shortage has attracted many foreign workers and by 1983 these foreigners made up 48 per cent of the workforce. About three-quarters of the foreign residents were male because the Libyan Government perceived men as most suitable for the type of work and the living conditions available. Thus the overall sex ratio of Libya in 1985, even after declining fortunes in the oil industry had led to the departure of many foreign workers, was 111.4 males per 100 females, compared to a ratio of 104.2 per 100 for the citizen population (Table 2.1).

Migration is a phenomenon associated with spatial differences in employment opportunities. Migrant workers, worldwide, come predominantly from countries which cannot find jobs for all their workforce at home. Examples of such 'labour reserves' are Botswana and Lesotho in southern Africa and the West Indies. These areas have feminine sex ratios with a ratio of 91 men per 100 women recorded for Botswana, and 93 for Lesotho (see Case study B) and Montserrat, a British colony in the Caribbean.

Many people left tiny Montserrat in the 1950s and 1960s to work in Britain. The 1960 census recorded only 80 men for every 100 women. For the age cohort over 70 years there were fewer than 40 men per 100 women although the sex ratio was masculine for the under-fifteens. Thus Montserrat society became predominantly one of grandmothers and children, with very few men of working age left behind on the island. After 1962 migration became more difficult because of legal

Plate 2.1 Migration to the colonization frontier. Sister and brother from Japan attracted by the opportunity for land ownership on a colony in Maranhao, Brazil. The woman is standing outside the house provided by the colonization agency. The traditional gender division of labour continues with the women still doing most household tasks, despite migration and unusual family structure
Source: The author

barriers introduced by the governments of the main receiving countries. Gradually Montserrat's prosperity improved as foreign residents and businesses were attracted by the stability offered by the island's colonial status. Many former migrants, having either reached retirement age or lost jobs because of recession overseas, decided to return to the land of their birth, and the island's population began to increase after a long period of decline.

Figure 2.4 shows the narrow-waisted population pyramid produced by these fluctuations in migration patterns. Birthrates were affected by the absence of people of reproductive age and fell from 29.5 per 1000 people in 1960 to a low of 17.7 in 1976 and then recovered to 22.3 in 1982. Mortality rates in the first year of life fell from 114.2 per 1000 live births in 1960 to only 7.7 in 1982. The island still has a high proportion of elderly females, now added to by return migrants and foreign retirees.

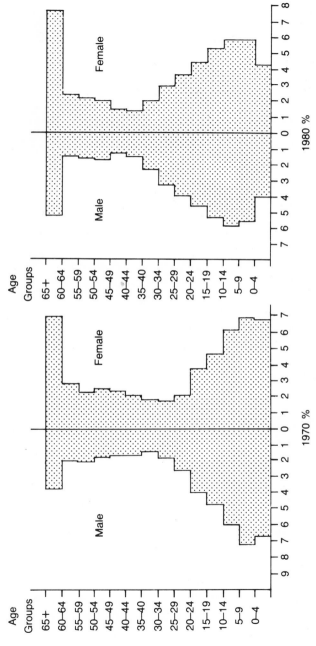

Figure 2.4 Montserrat, West Indies: age and sex structure, 1970 and 1980

In 1980, Montserrat had a sex ratio of 92.7 men per 100 women which was the most masculine ratio recorded in any census since 1871. The long history of female numerical dominance on this island has contributed to women's economic importance and independence. In 1972 women operated 44 per cent of small farms on Montserrat but this had fallen to only 23 per cent in 1983, as male return migrants replaced female farmers and women took advantage of better-paid employment elsewhere in the economy. At the same time, women managed to retain their dominance of the prestigous jobs in the civil service and local financial sector which they had moved into during the period of mainly male out-migration. Montserratian men explain this by relying on the now fallacious argument that there are more women than men of a working age on the island. Women were also able to continue to take advantage of the universal, free childcare which the government had been forced to introduce when there were few men available for the workforce.

Both men and women migrate but the reasons for the migration, the type of destination and the length of time spent at the destination are often gender specific. In so far as any general patterns can be identified men are more likely than women to migrate in order to gain educational qualifications while women are more likely to migrate to marry or to rejoin a migrant spouse, but autonomous female migration is increasing in importance, especially among younger women (see Table B.1). Migrant women may also be flouting traditional restrictions and norms. They may be avoiding arranged marriages, leaving a marriage that is unhappy or has not produced children, or escaping from low economic and social status.

Migration for both men and women may be short-term or circular rather than permanent and this temporal pattern will affect both the source region as well as the adaptation of the migrant to the receiving area. Remittances to family left behind are most consistent from migrants intending to return and regular visits by migrants bring new ideas into traditional rural areas. Teenage Indian women from the highlands of Peru are often sent to the cities to work as servants but are expected to return to their villages to marry. In Indonesia both men and women move between rural and urban areas in a circular manner responding to gender-specific labour demands in the countryside during the agricultural year.

Rural to urban migration involves the largest number of people but movement may also be from rural to rural areas or across international boundaries. Three factors affect female rural to urban mobility: female participation in agriculture, availability of economic opportunities for women in the cities and socio-cultural restrictions on the independent mobility of women. Internal migration from rural to urban areas is dominated by women in Latin America and parts of Southeast Asia and by men in Africa, South Asia and the Middle East, reflecting regional differences in the gender-specific pattern of labour demand. As a consequence of migration, the sex ratio for Latin American cities in the period 1965–75 was 109 women for every 100 men while in African towns the ratio was 92 women per 100 men. In colonial Africa women were discouraged from migrating to the towns and in Uganda in the 1950s all single women in Kampala were considered to be prostitutes and were by law repatriated to the countryside. Today rural poverty and backbreaking farmwork is driving women to the cities where they can find opportunities for education and economic independence.

Age, marital status and education of female migrants influence their employment status in urban areas. For young, poorly-educated, unmarried women domestic service, prostitition or, especially in Southeast Asia, factory work are the main jobs while older women work as smallscale traders, craftworkers or producers of beverages and cooked foods. In the Middle East educated women may find jobs in segregated all-female hospitals or educational establishments. Married women who move with their husbands to cities are less likely to seek employment.

International migration

Men tend to migrate over longer distances and to participate in international migration more than women. Males exceeded females in 83 per cent of all annual international migration flows between 1967 and 1976, with the exception of movement to the United States where, since 1930, women have constituted a majority of the foreign-born legal immigrants (61 per cent of those admitted between 1952 and 1978), and Asia, Latin America and the Caribbean have replaced Canada and Western Europe as the leading sources of immigrants. The Caribbean, the Philippines, Thailand and Turkey provide significant flows of autonomous women migrants across international boundaries.

Men and women migrants compete in separate labour markets. Among the vast majority of migrants who are poor and unskilled, men find a great variety of job opportunities available but women migrants tend to be concentrated in the garment industry, or poorly-paid service sector jobs (see Table B.2). Government policy in receiving countries causes changes in the sex ratios of migrants over time, depending on whether it is aimed at worker recruitment (and what type of work) or at family reunification.

The de-skilling of industry in the developed world has created a demand for people willing to work long hours at boring, monotonous jobs for low wages and so provided a niche for immigrant women. Many of these women enter on restricted permits linked to work for one employer or to the legal status of their husband. If their marriage breaks down or they lose their original job they may be deported. In this situation, immigrant women become the most exploited workers. They cannot complain if they have to work for less than the minimum wage and for very long hours. They often find jobs in hotels, restaurants, nursing homes, domestic service or in the garment industry as out-workers or in small sweatshops.

Another opportunity has arisen as a consequence of the increased proportion of employed married women in the richer countries which has expanded demand for domestic servants. Thus Canada supported a programme to bring in West Indian women to work as domestics, Sri Lankan women are sent to the Middle East, and migration from the Philippines to the United States, Britain and Singapore is female-dominated reflecting active recruitment of domestic workers. In the mid-1970s 57 per cent of all long-term work permits issued by Britain to Filipinos were for domestic work. By 1989 there were nearly half a million Filipinos working abroad and their remittances to their families back home accounted for more invisible earnings than any other sector of the Philippine economy. The economic importance of these migrants has prevented their home governments from enforcing regulations to improve the conditions under which they are hired.

Effects of out-migration on rural areas

When men migrate, leaving their wives and families behind in rural areas, the rural economy is affected. In Kenya and Zimbabwe two-fifths of rural families are headed by women and these women have a heavy burden of work leaving little time for leisure. Often African migrant husbands do not relinquish their decision-making authority in their household or native village, leading to delays in the implementation of community projects and a situation in which wives, who are expected to look after the cattle, may not sell or slaughter a beast without their husband's permission. The men have little incentive to use the land more efficiently as most of their income comes from the town. They see the land as a cheap place in which to raise their children and somewhere to retire. But, on the other hand, wages sent home by men working in the cities enabled many of these families to survive the drought of the mid-1980s and research in Zimbabwe showed that households receiving cash from a migrant earned a third more from farming than those without remittances because they were able to buy modern inputs such as fertilizer.

Case study B

Sex-specific migration and its effects on Lesotho

The women of Lesotho are gold widows: left behind to make of life what they can while the men work in the gold mines of South Africa. The border between the two countries forms a sex-specific barrier as, since the 1960s, only men have been allowed to enter South Africa to work. Lesotho society is, therefore, one in which women do most of the work of agriculture and social reproduction. Yet women are generally better educated than men and have developed considerable autonomy. This leads to a deep sense of frustration among women because they are denied access to the modern industrial world. All that most of them can now do is to remain on South Africa's periphery, reproducing its labour force, doing unpaid domestic work, cultivating infertile soil, seeking low-paid local employment, providing a market for South Africa's goods and becoming increasingly dependent on the unreliable supply of remittances from male wage earners.

Traditionally women have moved to their husband's home on marriage so they have dominated the pattern of rural to rural moves while men have played the major role in international migration. Now women have started to move to the town and in a survey of young urban migrants it was found that the sex ratio of migrants under the age of 24 years was 49 males per 100 females (Table B.1). Women are moving urbanwards because life in the rural areas has become intolerable. About two-thirds of rural households are headed by women but only about half of these have migrant husbands and receive remittances; the rest are widows, reflecting the high rate of fatal accidents in South African mines.

Case study B *(continued)*

Table B.1 Maseru City, Lesotho: age of migrants at time of move, by sex

		Age			
	Under 24	*25–34*	*35–44*	*Over 44*	*% of total*
% of men	32.8	32.8	23.0	11.5	45.3
% of women	55.7	24.9	12.2	7.2	54.7

Source: Wilkinson (1985)

The well-educated young women of Lesotho are faced with a situation in which South African 'influx control measures' keep them penned within their impoverished little country where they are further controlled by a traditional, patriarchal society. They see Maseru, the capital city, as providing them with some hope of freedom, and in 1978 43 per cent of female migrants declared that they would never return to their rural homes, compared with only 30 per cent of the men. Yet employment is often illusory. A survey of *de facto* household heads in the city showed that 53 per cent of the women but only 17 per cent of the men were unwaged. Consequently large numbers of women are forced to take up petty trading, the illegal sale of home-brewed beer and prostitution. Where jobs in the formal economy are obtained by women, they are predominantly in the low-paid clerical and domestic service sectors (Table B.2).

Table B.2 Occupations of migrants to Maseru City, Lesotho

Occupations	*Men*	*Women*
	%	
	N=416	
Labouring	9.0	0.0
Construction	21.8	0.4
Engineers and drivers	13.3	0.4
Clerical	10.1	13.2
Sales	6.9	3.5
Professional and managerial	14.3	7.0
Domestic	1.6	19.7
Others	7.5	4.4
Unemployed	15.4	51.3

Source: Wilkinson (1985)

Case study B *(continued)*

> Prostitution is one of the better-paid jobs. In its modern form it is associated with the growth along the border of Las-Vegas style entertainment centres catering for white males from South Africa. Weekend trips for this purpose have become part of Lesotho's tourist trade. Thus the border is permeable in both directions by only one sex. Lesotho men serve South Africa in her mines while Lesotho women are forced to stay at home and educate themselves to provide services for white South African males, services which are illegal within South Africa.
>
> *Source:* C. R. Wilkinson (1985) 'Migration in Lesotho: a study of population movements in a labour reserve economy', unpublished Ph.D. thesis, University of Newcastle upon Tyne, England

Without remittances households headed by single mothers in rural areas are significantly poorer than male-headed households. Men farmers in Botswana in the 1970s were twice as likely to own the cattle needed for ploughing, milk and financial security than women farmers and their crop yield was generally four times greater. In colonization zones of eastern Colombia abandoned wives find it very difficult to assume the role of farmer even though their children are hungry. In the Caribbean, despite a tradition of female agricultural work, male migration also causes problems. Women's shortage of time and difficulties in obtaining assistance with farm tasks considered to be male has led to a decline in agricultural output and underutilization of land. Thus the feminization of agriculture is usually accompanied by increased poverty and malnutrition among rural families. More West Indian women than men farmers are dependent on remittances from migrant relatives but these funds are seldom invested in agriculture. Instead they are used to improve rural living conditions and to finance migration for other members of the family.

Female-headed households

Labour reserves export their excess male labour and are left with a society made up of families headed by women. In the Caribbean about one-third of household heads are women. This proportion ranges from

50 per cent in St Kitts and 44 per cent in Montserrat to less than 20 per cent in Guyana. In Brazil, although the total for the country as a whole is only around 15 per cent, spatial variation is also marked: female-headed households are most common in the very migration-prone, arid northeast and in urban rather than rural areas. For the Third World as a whole, estimates vary between one-sixth and one-third of all households but everywhere it is seen as a phenomenon which is increasing rapidly with modernization.

Female-headed households may be the result of the breakdown of male-headed households through death, marital instability or migration. They may also occur in a situation where the woman has no permanent partner or when the husband has several wives. Regional patterns are distinguishable: in Asia widowhood is still a prime cause; in North and southern Africa and the Middle East international migration is the predominant reason; in West and Central Africa male migration to cities leaves women alone in rural areas; in the Caribbean many women choose to have short-term visiting relationships but no permanent resident partner and this pattern is exacerbated by international migration.

In societies where property is corporately held and the household is the unit of labour, women rarely emerge as heads of households. Female-headed households will develop where women have independent access to subsistence opportunities through work, inheritance, or state-provided welfare and are permitted to control property and have a separate residence. Their subsistence opportunities must be reconcilable with childcare and must provide an income not markedly lower than that of men of the same class. Development has been accompanied by increased privatization of the means of production and a decline in cooperation within kin groups and has thus provided the conditions for the growth of female-headed households.

These households are often among the poorest as they contain fewer working adults than male-headed households and women earn lower wages than men. Their composition has also been said to constitute a poverty trap with children disadvantaged because they may have to leave school early to seek paid employment or take over household chores to allow the mother to work outside the home. Maternal neglect and lack of paternal discipline has been thought to encourage truancy and delinquency and to perpetuate a familial pattern of deprivation. However, households headed by women are not undifferentiated and should not necessarily be seen as victims of development. In some cases women choose to establish their own household in order to gain decision-making independence and to escape male violence and economic reliance on an irresponsible man. Such households have a positive effect on female autonomy and despite suffering from stigmatization as a deviant form, many function very successfully both socially and economically.

Key ideas

1 Most countries of the world have a feminine sex ratio because women normally live longer than men.
2 The highest proportions of women are in labour reserve areas from which many men have migrated.
3 Masculine sex ratios occur, and in many cases are increasing, in South Asia, China, North Africa and the Middle East.
4 High masculine sex ratios are seen in areas in which women have little economic value as producers and low social status.
5 More men than women migrate across international boundaries but within states migration to cities is predominantly female in Latin America and male in most of Africa.
6 Female-headed households are increasing rapidly as a result of sex-specific migration, changing attitudes to marriage and declining support for single women from family and community.

3
Reproduction

The term 'reproduction' is a chaotic concept which not only refers to biological reproduction but also includes the social reproduction of the family. Biological reproduction encompasses childbearing and early nurturing of infants, which only women are physiologically capable of performing. By social reproduction is meant the care and maintenance of the household. This involves a wide range of tasks related to housework, food preparation and care for the sick, which are usually more time-consuming in the Third World than in the industrialized world. In most countries women are also expected to ensure the reproduction of the labour force by assuming responsibility for the health, education and socialization of children. Third World countries generally offer less state assistance for these tasks than is provided in First and Second World countries.

In addition to household maintenance, social reproduction also includes social management. This latter role of women is often ignored. It involves maintaining kinship linkages, developing neighbourhood networks and carrying out religious, ceremonial and social obligations in the community. The survival strategies of many poor women depend on their success in this role. Local and kin groups can help when members of the family become ill, need a job or a loan or are faced with some other sort of crisis. A woman's success as a social manager may bring status to her family and to herself and enable her to take on leadership positions within the community.

Reproduction may be distinguished from production on the basis of the law of value. Reproductive labour has use-value and furnishes family subsistence needs while productive labour generates exchange-values, usually cash income. Empirically this separation is very difficult to make as, within the domestic sphere in which most women work, both categories of tasks are interrelated and enmeshed in a totality of female chores. Any one task may have both use- and exchange-value at different points in time. Yet it is analytically useful to accept this division as a theoretical framework within which to consider the diversity of women's domestic labour.

It is increasingly being realized that the task of reproduction is a major determinant of women's position in the labour market, the sexual division of labour and the subordination of women. The household is the locus of reproduction so that social relations within the household play a crucial role in determining women's role in economic development.

With modernization and industrialization, unpaid housework becomes increasingly isolated and spatially separated from paid productive work outside the home. Women's participation in the productive labour force will inevitably be affected by the time and energy burden of their reproductive tasks as well as by the power relationships between household members. Large families can be seen as an opportunity cost for women, limiting life choices. In order to understand fully the nature of women's subordination and their role in the development process, it is essential to study both reproduction and production and the inter-relations between them.

Friedrich Engels (1820–95), a close associate of Karl Marx, saw reproduction as the key to the origin of women's subordination by men. He believed that it was associated with the introduction of the concept of private property. The wish of the property owner to pass his property on to his children led to the need to identify the paternity of these heirs by controlling women's sexuality, and then to ensure their survival by regulating her reproductive activities. However, Engels assumed that woman's participation in productive activities, as a result of the spread of industrialization, was a necessary precondition for her emancipation. It is now clear that women's increasing involvement in the wage economy in the developing world has not ended their subordination. Rather, it has been accompanied by the transfer of patriarchal attitudes from the household to the factory, and the desire to seclude women within the family has encouraged outworking in the home at very low wages. Development has not brought greater freedom for women and in many cases women are now expected to carry the double burden of both reproductive and productive tasks.

Women in the Middle East and North Africa have the lowest rates of economic activity and this is linked to Islamic patterns of female control. Marriage is compulsory for the faithful but is legally an unequal institution: men may have up to four wives and infertility or failure to bear sons are grounds for divorce. In many Muslim countries women are not only segregated from men but have seclusion or *purdah* imposed on them and have to wear long, concealing garments and sometimes a veil in public. Female circumcision, or genital mutilation of varying degrees of severity, is associated with, although not limited to, Muslim countries. It is believed to ensure the conjugal fidelity of women by destroying their ability to enjoy sex. It is usually carried out by female relatives on girls between the ages of 5 and 8 years and causes many medical and psychological problems. Despite a growing recognition of the problems, many mothers still feel it is essential in order to ensure their daughter's chances of making a successful marriage, and so it still occurs even where it has been made illegal.

Biological reproduction

Fertility, that is the total number of children born, on average, to each woman during her reproductive years, is probably the best documented aspect of women's lives, for reasons which would have been clear to Engels. On the whole, fertility rates in the Third World are much higher than in the First and Second Worlds. At any one time it is thought that one-third of Third World women are either pregnant or lactating. In the Gambia the average woman has 10 to 14 complete pregnancies and spends virtually all her reproductive years either carrying a child in her womb or breastfeeding a baby. The physiological stress of this reproductive activity and its effect on the woman's ability to undertake tasks related to household maintenance, such as collection of water, fuel gathering, food processing and subsistence farming, need to be considered in development planning.

However, within the Third World fertility rates vary enormously from country to country. The city-state of Singapore with an average rate of only 1.8 children per mother, a level equal to that of the United Kingdom, may be contrasted, at the other end of the scale, with Zimbabwe which has a rate of 8.0 children per mother. In Africa and the Middle East the figure is above 5 except in Gabon. Total fertility rates have been falling: between 1950 and 1980 worldwide rates fell from 4.9 to 3.6 children per woman and the rate for the developing world fell from 5.9 to 4.1. The decline was most marked in the Far East where the fertility rate fell from 5.5 in 1950 to 2.9 in 1980, strongly influenced by China's one child policy. In sub-Saharan Africa, however, the total fertility rate over this same period rose from 6.4 to 6.6 as infant mortality rates declined and male migration to cities left women increasingly dependent on children's labour in subsistence farming.

It is thought that even today two-thirds of all couples in the Third World do not have access to contraception, resulting in many unnecessary deaths of women from illegal abortions, and too frequent, early and poorly spaced pregnancies. Clearly fertility rates are related to levels of development although it would be wrong to assume that large families are always considered negatively by women. As women move to cities, become better educated and find new opportunities for work and self-development outside the home, the birthrate tends to fall. In cities children are less useful as supplemental labour and are more costly to maintain. Spatial differences in fertility may also reflect the different sex ratios of rural and urban populations and the distribution of female-headed households.

The spatial effects of these changes can be seen clearly in Peru. In the capital city of Lima, which as the nation's main focus of modernization and employment opportunities is attractive to women migrants, the fertility rate is 3.5. In the rural areas of the Andes, where many women are illiterate, do not speak Spanish and have no access to contraceptives, the fertility rate is 8.1. Yet despite this huge regional disparity in actual fertility rates, women in both rural and urban areas of Peru declare that their desired number of children is only three. This gap between hope and reality indicates enormous unmet demand for access to means and opportunities that would make possible a reduction in birthrates.

Both knowledge and place of residence influence implementation of family planning. In Jordan it was found that 99 per cent of urban, 98 per cent of semi-urban and 89 per cent of rural potential users were aware of the utility of contraception but only 6 per cent in rural areas, 24 per cent in semi-urban and 38 per cent in urban areas actually utilized it. In Africa the variation in usage in 1985 ranged from 5 per cent in Lesotho and 6 per cent in Nigeria to 29 per cent in Zimbabwe and 53 per cent in Mauritius. In many Third World countries surgical methods of contraception are popular as the regular use of less permanent forms is difficult in crowded living conditions without privacy and running water. Spatial variations in the methods and levels of usage of contraception have been linked to accessibility and awareness of family planning facilities, to the relative costs and benefits of raising children, and to attitudes towards the quality of life desired for one's children.

It is generally assumed that with development the trend in the implementation of family planning is ever upward. A study among women market traders in the city of Ibadan, Nigeria, illustrates the rapidity with which both increases and decreases can occur. Although there were consistent differences related to religion and education with 44 per cent of Christians but only 23 per cent of Muslims, and 45 per cent of literate women but only 23 per cent of illiterates using contraception, usage levels for all women fluctuated with the economic health of the nation, declining during the oil boom of 1974 but increasing with the economic uncertainty and high inflation rates of the 1980s.

National policies also affect fertility rates. Libya's need to increase her labour force has led to the encouragement of large families. Singapore's introduction of fiscal incentives to encourage educated women to have more children resulted in an immediate increase in births. On the other hand, China and India, faced with huge and rapidly expanding populations, opted for direct state intervention to reduce the birthrate. The setting of local quotas for male sterilization in India, reinforced by bribes, eventually made the programme politically unacceptable. China's enforcement of a policy of only one child per family, in a culture which has traditionally given more status to male children, resulted in increases in female infanticide, and by 1988 there was evidence of the use of sex-selective abortion after sex-determination tests to reduce the number of girls born. Such excesses and the emergence of a generation dominated by overly-pampered little boys eventually led to a relaxation of the official policy in rural areas and for minority ethnic groups. In Bangladesh the legal age for marriage for women was raised to 16 in order to lower the fertility rate and reduce the number of maternal deaths due to early and too frequent childbirth. However, in rural areas few people are aware of this law and most women get married between the ages of 12 and 14 years.

Cultural and religious attitudes to large families, particularly where male status is strongly linked to a man's ability to father many, preferably male, children, may counteract other pressures for a reduction in the fertility rate. When economic pressures force women to depend on elder daughters to take on household maintenance tasks in order to free the mother for wage earning outside the home, then female children may be preferred. Where these countervailing pressures are strongest, or female subordination in marriage most acute, women will seek out contraceptive advice secretly or may reject marriage.

In southern China in the nineteenth and early twentieth centuries there was a strong feudal and patriarchal society. Footbinding, concubinage, prostitution, female infanticide and sale of females were all institutionalized and reinforced by powerful ideological mechanisms. At times of economic adversity practices such as the sale of females into prostitution and domestic slavery were intensified and still continue today. In the silk producing and rice growing areas women played a vital economic role and this gave them the independence to resist marriage. Anti-marriage groups were set up because women both feared the subordination of marriage and enjoyed the economic independence and friendship of other women provided by work. Typically they formed sisterhoods and lived in all-female households. Some who were married paid for replacement concubines while other sisterhoods banded together to prevent their husbands taking concubines. The anti-marriage movement was strengthened by high male migration in the early twentieth century when single women became the sole support of many peasant households. Some of these women adopted 'daughters' and the groups organized savings schemes and retirement homes. In the late 1920s depression in the silk industry forced many of these women to migrate. Some went to Malaya where they set up similar anti-marriage sisterhoods and worked in tinmining, domestic service and plantation rubber production.

Case study C

The unengaging impact of education in Singapore

The Social Development Unit (SDU) of Singapore is a state-run marriage bureau initially set up to find partners for women graduates. The need for such an organization first became apparent in 1982 when ministers studied the returns of the 1980 census. They found that a large proportion of female graduates were not married. "Being government people they did not know what to do except make speeches," says Dr Eileen Aw, who now heads the agency from the 40th floor of the Singapore Treasury

Case study C *(continued)*

building. Finally they sent a team to Japan which came back with the idea of a marriage bureau.

Dr Aw was asked to run it when it was set up in 1984. Brusque and lively, she says that before being approached she had never imagined doing such a job. But though she was surprised by the offer, she was also aware that the traditional Chinese males' habit of preferring wives less educated than themselves was creating difficulties for a new generation of women who had graduated from university. "I was conscious that some of my educated female friends were interested in getting married but had not found husbands," she says. "I thus saw there was a problem."

In 1984 the National University of Singapore adjusted its entrance requirements to make it easier for men to get in. Men had found it more difficult than women to meet the second language requirement and the sex ratio of students had begun to swing in favour of women. Since 1979 the university has limited the number of women admitted to the prestigious medical faculty to one-third of each year's intake. While Singapore men want to marry women educationally below them, Singapore women will not accept men of inferior education and so 39 per cent of female graduates remain single. The result is not only that a large number of female graduates are unmarried but also that 38 per cent of men without higher education fail to get married. For a Singapore government sensitive to fertility rates among a small population, the further worrying conclusion was that some of the island's brightest women were failing to reproduce.

The task of the SDU is to bring together single male and female graduates in order to enhance their opportunities of finding a mate. In its first three years of operation the SDU succeeded in finding marriage partners for 400 graduates. The agency fills a gap created by the pressures of life in modern Singapore. Jobs are markedly gender segregated so that, for example, teachers tend to be women, and engineers tend to be men, with very little opportunity for meeting each other. In addition, working men and women return home so tired after what often amounts to a 12-hour day, that they have no time to plan social activities in the evening.

Case study C *(continued)*

> In Singapore marriages used to be arranged. Couples no longer want that, but they still retain some of the attitudes of an earlier time. Dr Aw believes that women have been a powerful force for social reform. They are more welcoming to Western ideas than men because they have an interest in breaking down the traditional, patriarchal values that for a long time kept women in an inferior position in Eastern societies. Men, by contrast, have shown their conservatism in their reluctance to amend the inheritance laws, to open up to women the same job and pay opportunities or to abandon those domestic privileges that work to their advantage.
>
> The subservient role of women, most noticeable in Japan, Korea and Taiwan, has its roots in Confucian culture. In Korea it is a tradition that when a girl marries and is about to live with her husband, her mother gives her a stone saying "even if your husband and mother-in-law provoke you, you only open your mouth when the stone starts to speak". A measure of women's new independence came in replies given to a recent survey of public opinion in Singapore. Ninety per cent of men said they thought marriage was necessary for a "full life". Only 80 per cent of women believed this to be the case.
>
> *Source: Financial Times* 30 June, 1988

Education

As Case study C shows, the educational levels of men and women affect many life options. In 1960 only 58 per cent as many women as men were literate in the developing world and by 1980 this figure had risen to only 74 per cent. Furthermore women's illiteracy is more easily concealed than men's and so it is probably underestimated. In general, Third World parents actively seek education for their children as the best means of improving their income-earning options, but over-burdened mothers may be forced to take daughters out of school to assist with childcare and household chores.

Figure 3.1 shows the international pattern of relative male and female literacy rates. In the industrialized world of the North there are virtually no gender differences in literacy and this is true also of South Africa and

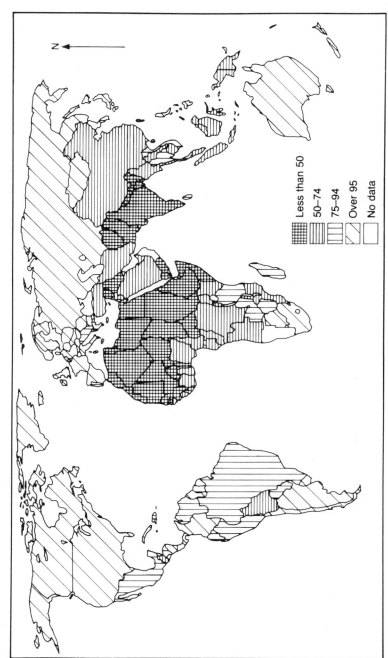

Figure 3.1 Female adult literacy rates as a percentage of male literacy

the Philippines and most of the countries of Latin America and the Caribbean. Yet female illiteracy rates of over 90 per cent are found in the countries of sahelian Africa, and in Morocco, Pakistan, Afghanistan, Nepal and Yemen. On the other hand, more men than women are illiterate in Uruguay, Bermuda, St Lucia, Botswana, Lesotho, São Tomé and the Seychelles, and in 1986 more women than men had completed secondary education in Botswana, Lesotho, Singapore, Sri Lanka, Chile, Venezuela, Panama, Barbados, Dominican Republic, Jamaica, and Trinidad and Tobago. Women achieve lower levels of education than men in the majority of Third World countries because of their confinement to the domestic sphere and male prejudice against educating women.

Women's participation in higher education in the Third World has increased very rapidly in the last two decades. Between 1965 and 1985 estimated female enrolment in tertiary education in South Yemen and Qatar increased from 0 to 19 and 57 per cent of the total respectively; in Guatemala from 9 to 28 per cent and in Brazil from 25 to 48 per cent. By 1985 women made up 45 per cent of students in tertiary education in Latin America, as they did in Europe, and 29 per cent in Africa. However, access to higher education is strongly dependent on class, location and income. Within the education system women tend to be channelled into certain subject ghettos such as nursing, education and social work, while the courses leading to the best-paid jobs, such as medicine, law and engineering, are still dominated by men.

Social reproduction

Activities carried out to maintain and care for family members are generally ignored in national accounts but they are essential economic functions which ensure the development and preservation of human capital for the household and for the nation. Education of young children and organization of the household so that members can maximize their access to educational facilities may be part of these activities. Others may include fuel and water collection, care of children, the sick and the elderly, washing clothes, and processing, preparing and cooking food. For most families throughout the world these jobs are done by women.

Much of the research focusing on Third World women has looked at biological reproduction in isolation from commodity production and has ignored social reproduction. Women perform the great bulk of domestic tasks in all societies. Even in Cuba where, by statute, men are supposed to assist women in such work, 82 per cent of women in the capital city, Havana, and 96 per cent of the women in the countryside have sole responsibility for domestic chores. The equivalent figure in Britain in 1988 was 72 per cent. In subsistence societies, the separation between reproductive and productive tasks is to a large extent artificial, symbolized by the woman with her baby on her back working in the fields (see Table 3.1).

Table 3.1 Sri Lanka: gender roles in household activities

Activity	% of total workhours spent on each activity	
	Male	Female
Food preparation	8	92
Winnowing and parboiling rice	0	100
Preserving food for the hungry season	20	80
Storing grain at harvest time	70	30
Production of fruits, tubers, greens and vegetables for home consumption	20	80
Fetching water	2	98
Collecting firewood	35	65
Upkeep of house and yard	5	95
Bringing up children	10	90
Bathing children	20	80
Attending to the sick in the family	15	85

Source: A. Wickramasinge 'Women's role in rural households and agriculture in Sri Lanka' in J. H. Momsen (ed.) *Geographical Studies in Women and Development,* (provisional title) to be published by Routledge

It has been estimated that in sub-Saharan Africa women spend an estimated four hours a day on collecting firewood and water, childcare and preparing food. This is in addition to the time they spend on agriculture, craftwork or trading. At certain times of year the domestic tasks will take longer than normal. In the dry season the village well may run dry and so women have to walk further for daily supplies of water (see Figure 3.2). If the nearby source of firewood is overutilized women will have to spend longer searching for fuel and perhaps also reduce the amount of cooked food that they prepare. At planting and harvesting times, when more time must be spent in the fields, domestic work must be cut down.

Families with several small children absorb much of women's time in childcare unless there are older siblings who can assist the mother,

Figure 3.2 Women's use of time
Source: The International Women's Tribune Centre

although she may not wish them to do so if it means giving up their opportunity to attend school. It has been shown that the nutritional level of children is often negatively related to the distance mothers have to walk to collect water. The average round trip from house to water supply in Africa is 5 kilometres and thus the effort of carrying water can absorb 25 per cent of a woman's calorific intake. Domestic chores in developing countries, where household appliances are rare, consume a high proportion of women's energy and time (Table 3.2). The recent introduction of village grain mills in the Gambia was found to save women 60 to 90 minutes per day. But it was the saving in energy that was most appreciated as the pounding of sorghum is hard work.

Housing

Housing conditions have an effect on the time and effort consumed in housework. In 1980 only a quarter of Third World urban dwellings had piped water and even fewer had sewerage systems. In rural areas the

Plate 3.1 Collecting water in India
Source: Janet Townsend, University of Durham

Plate 3.2 Women buying and selling in the market in Merida, Yucatan, Mexico
Source: The author

Table 3.2 Time and effort expended on selected domestic tasks in Keneba, The Gambia

Activity	Total time (minutes per day)	% time spent on hard work	moderate work	light work
Cooking:				
Boiled staple and sauce	94.0	2.1	48.9	48.9
Preparing breakfast	16.5	0.0	84.8	15.2
Pounding:				
Millet/sorghum, off the stem	117.0	36.8	34.2	29.0
Millet/sorghum, bran removed	44.0	56.8	18.2	25.0
Millet/sorghum, endosperm powdered	66.0	42.4	15.2	42.4
Rice, husk and bran removed	51.0	51.0	25.5	23.5
Drawing water from well	49.0	14.3	44.9	40.8
Sweeping:				
Compound	32.0	0.0	96.9	3.1
House and kitchen	10.0	0.0	100.0	0.0
Washing:				
Clothes at well	180.0	0.0	89.4	10.6
Bowls and pots for one meal	13.0	0.0	100.0	0.0

Source: H. R. Barrett and A. W. Browne (1989) 'Time for development? The case of women's horticultural schemes in rural Gambia'. *Scottish Geographical Magazine*, 105(1) p. 5

situation was worse. A wooden shack with an earthen floor suffers from dust in the dry season and mud in the wet season. Cleaning is never ending and it is difficult to see any positive results. Shanty towns without electricity, piped water, paved roads or sewers make both housework and childcare harder and more time-consuming (Table 3.3). It is not surprising that in many Latin American cities women have mobilized and led local protest groups demanding better access to services for shanty town dwellers.

Table 3.3 Hours spent by women each day on housework according to family structure and settlement level of service in Queretaro, Mexico

Family type	Settlement quality Best	Intermediate	Worst	Mean hours
Nuclear	8.6	10.7	11.4	10.2
Extended	8.4	9.2	10.1	9.2
Mean hours	8.5	9.9	10.7	

Source: Sylvia Chant (1984) 'Women and housing: a study of household labour in Queretaro, Mexico' in J. H. Momsen and J. Townsend (eds.) *Women's Role in Changing the Face of the Developing World*, Women and Geography Study Group of the Institute of British Geographers, Durham University

Overcrowded housing conditions encourage the spread of infectious diseases and the lack of privacy may cause psychological stress and family tensions. Thatched roofs, and dirt walls and floors provide refuges for disease-carrying insects and for poisonous snakes and centipedes. Poorly-ventilated housing may be associated with respiratory infections and cooking indoors over smoky open fires contributes to chronic lung diseases such as bronchitis.

The household

The size and make-up of the household determines to a large degree the burden of work on women. It has been shown that in nuclear families the full burden of social reproduction falls on the wife and mother, but in extended and female-headed households there is much more sharing of tasks because the mother has more autonomy (see Table 3.3). In the developing world the nature of households is changing very rapidly. Co-residential households may not necessarily be child-rearing units, nor are they always economic units and the role of non-residential migrant members may be crucial to an understanding of the function of the household.

It is usually assumed that the household head is male and that he allocates household labour and organizes the distribution of consumption goods among household members so that all benefit and participate equally. Clearly this is not always so and much depends on gender relations and power within the household. In general, women's duties are closely associated with the collective aspects of family consumption while men have more individual control over their own personal consumption of resources. In order to understand the role of the household in development it is essential to recognize the dynamics of the system: the changing nature of production, distribution and consumption relations within the household, and the effect of life cycle on dependency ratios.

Key ideas

1 Fertility declines with urbanization and education of women. This decline is helped by improved availability of contraception and acceptance of its use by both men and women.
2 More men than women are literate, especially in Africa and Asia.
3 Most women work longer hours and have less leisure than men.

4 Household structure is becoming more complex and the proportion of nuclear households increases with modernization.
5 Type of family, age of children, seasonal shortages and work pressures, as well as housing conditions, affect the time and energy women need for social reproduction.

4
Women and work in rural areas

One of the major demographic trends in the Third World is the movement of people from the countryside to the cities (see *Population and Development in the Third World* by Allan and Anne Findlay in this series) but this movement is sex-specific and the predominant group varies from country to country. Figure 4.1 shows clear spatial differences in rural sex ratios at the continental scale. In Latin America more women than men move to cities leaving a masculine sex ratio in the rural population, except in Bolivia. In South Asia and China the nation-wide masculine sex ratios are most marked in the rural areas. In Africa and in the Middle East migration to cities is predominantly male and a higher proportion of women stay in the countryside. Rural to rural population movement also occurs. Where this is to a frontier of settlement as in Amazonia, then families generally move as a unit, but eventually women and children may be left alone on their holding while men seek paid employment elsewhere.

Yet if we change the scale of analysis we can see that the broad picture of rural sex ratios at the continental level obscures some very important regional variations. For example, men migrate from the rural areas of the highlands of Peru to jobs in coastal fishing or to take advantage of opportunities offered by the colonization of the lowland rainforest areas of the interior, leaving women and children behind. In India, despite a very marked masculine rural sex ratio nationally, the southern state of Kerala had an overall female sex ratio of 101.6 women per 100 men in 1971, a ratio which rose to 102.7 women per 100 men in

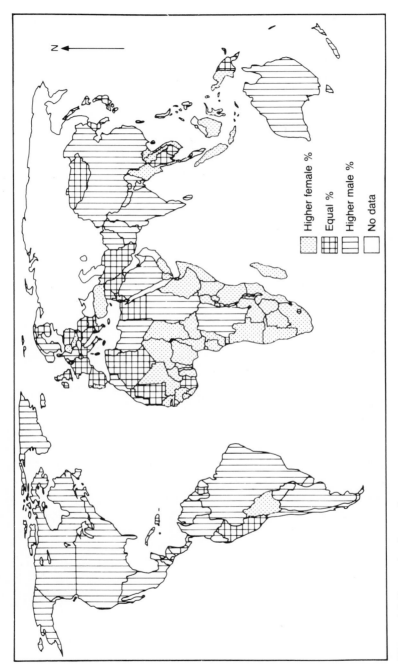

Figure 4.1 Sex ratios in rural areas

Higher female %

Equal %

Higher male %

No data

Plate 4.1 Women in paddy fields, India
Source: Janet Townsend, University of Durham

Plate 4.2 Brazilian female and male labourers picking black pepper on Japanese owned farm in Para, Brazil
Source: The author

lowland rural districts. These spatial variations in the sex ratio reflect location-specific differences in the economic role of women in rural areas.

Some 70 per cent of women living in rural areas of Third World countries work on the land. Table 1.1 showed that women are most likely to be agriculturists in the poorest countries but there are very distinct variations at the continental scale. The female participation rate in the agricultural labour force is highest in sub-Saharan Africa, Asia and the Caribbean and lowest in Latin America. However, the apparently minor role played by women in Latin American agriculture may be changing. Figures for Brazil show that between 1970 and 1980 the total number of men economically active in agriculture fell, but the number of women increased and the female proportion of the farm labour force rose from 9.6 per cent to 12.7 per cent. Similarly, in Peru, the proportion of the female population which was economically active in rural areas grew from 14 per cent to 21 per cent between 1972 and 1981. It is clear that the widespread poor reliability of statistics on female employment, particularly in agriculture, make it impossible to discern a trend of declining female participation in agriculture with development, in line with the well-established general relationship. Indeed the existing data appear to suggest the opposite.

Statistical evidence on gender roles in agriculture is very unreliable. In many societies it is culturally unacceptable both for a woman to say that she works in agriculture and for the census taker to consider that she might have an economic role. Detailed fieldwork has often indicated a much higher level of female participation in agriculture than is generally recorded in national censuses. Official classifications may also make it difficult to understand the role of women and to make cross-national comparisons. Changes in employment status, from independent cultivator to unpaid family worker with the expansion of cash cropping in Africa, from independent cultivator to wage labourer in India as landlessness increases, and from permanent hacienda worker to wage-earning rural proletariat in Latin America with the rise of agribusiness, appear to be disproportionate among women workers. Some of these variations in the recording of the role of women in national censuses may reflect societal changes in the perception of women's roles.

Farming systems and gender roles

It has been argued that regional differences in the female contribution to the agricultural sector are related to the way in which people extract

Plate 4.3 Aborigine women hunting, Australia
Source: Elspeth Young of the Australian Defence Force Academy, Campbell ACT

Plate 4.4 Aborigine women gathering, Australia
Source: Elspeth Young of the Australian Defence Force Academy, Campbell ACT

food from the environment. Before we became agriculturalists some 10,000 years ago we were all hunters and gatherers and it is thought that women produced the major part of the food consumed. Studies of the few remaining hunting and gathering societies, none of which is totally untouched by the modern world, have shown that gathering by women is the major source of food in over half of those societies. Women may also participate in communal drives of herds, may be responsible for hunting small animals and collecting insects and reptiles, and may work with men in fishing. Hunting is precarious and uncertain and so gathering tends to provide the basic diet. In contemporary hunting societies like those of many aboriginal groups in Australia, the nutritious grubs and 'bush tucker' gathered by the women often provide a major part of the food intake, but only the meat which the men procure is socially valued and shared.

Women may have been the first agriculturists, as the step from gathering roots and seeds to planting and cultivating is but a small one. As agriculture has developed it is possible to recognize male and female farming systems. In the extensive, shifting, non-plough agriculture of tropical Africa and South and Southeast Asia most of the work in the fields is done by women and the system is deemed to be female. In the plough culture of Latin America and Arab countries there is low female participation and the system is identified as male. Both men and women are equally involved in the intensive irrigated agriculture of Southeast Asia. This explanation of gender roles is attractive but easily becomes a form of agricultural determinism. At the local level relationships between types of farming system and female participation are very complex.

Farm size

Geographical region appears to be a major explanatory variable in differentiating women's share in Third World agriculture, but the distribution of land between large and small farms proves to be more important. Areas with many small farms usually have relatively high proportions of women agricultural workers. Land distribution accounts for 44 per cent of the total variation in the female share of farm labour and no less than 80 per cent of the variation attributable to regional differences. The highest proportions of smallholdings are found in sub-Saharan Africa, Asia and the Caribbean while Latin America has the lowest.

In general, farms run by women tend to have poorer soil and to be smaller and more isolated than those cultivated by men. The crop–livestock mix on female-operated farms is also different, with the emphasis on production for home use rather than for sale, and where cash production does occur, sales are made predominantly in local markets rather than for export. Small animals such as chickens and pigs, which can be fed on household scraps, are kept more often than cattle. Because women smallholders often find it difficult to hire men to undertake tasks such as land preparation and pesticide application, they may be forced to leave some of their land uncultivated and to concentrate on subsistence production.

Gender divisions of labour in agriculture

The particular tasks done on farms by men and women have certain common patterns. In general, men undertake the heavy physical labour of land preparation and jobs which are specific to distant locations, such as livestock herding, while women carry out the repetitious, time consuming tasks like weeding, and those which are located close to home, such as care of the kitchen garden. In most cultures the application of pesticides is considered a male task as women are aware of the danger to their unborn children of exposure to chemicals. Women do a major part of the planting and weeding of crops. Care of livestock is shared, with men looking after the larger animals and women the smaller ones. Marketing is often seen as a female task, although men are most likely to negotiate the sale of export crops. Children may assist in feeding and herding livestock. This division of labour is not immutable and may be overridden by cultural taboos. Some tasks are gender-neutral. The introduction of a new tool may cause a particular job to be reassigned to the opposite sex and men tend to assume tasks that become mechanized. Behaviour in the individual household is often more flexible than the broad general picture suggests because of personal preferences and skills, economic necessity, or the absence of key members of the family.

In areas with high male migration many women become farm operators in their own right. In the Eastern Caribbean 35 per cent of small farms are run by women. Table 4.1 shows the gender division of farmwork on two islands: Montserrat, with an Afro-Caribbean population, has a relatively high proportion of female farmers while Trinidad, with a majority of East Indian small farmers, has very few women farm operators. These differences have resulted in less gender-specificity of tasks in Montserrat than in Trinidad although farm tasks appear to be shared to a greater extent in Trinidad.

Table 4.1 Gender division of labour on small farms in Trinidad and Montserrat, West Indies

Task	Trinidad			Montserrat		
			% of farms			
	Male	Female	Joint	Male	Female	Joint
Preparation of soil	100	0	0	65	20	15
Planting	72	14	14	42	38	20
Weeding	0	50	50	35	42	23
Pest control	84	6	10	75	25	0
Fertilizing	34	33	33	75	25	0
Harvesting	16	34	50	31	45	24
Care of livestock	14	49	37	53	16	31

Source: I. S. Harry (1980) 'Women in agriculture in Trinidad', unpublished M.Sc. thesis, University of Calgary, Canada; and J. H. Momsen (1973) fieldwork

Women and agricultural development

The impact on women of the modernization of agriculture is both complex and contradictory. It varies according to the crops produced, the size of farm and the farming system, the economic position of an individual farm family and the political and cultural structure of societies. Table 4.2 indicates some of the possible effects on women of a number of different changes in rural areas.

Women have often been excluded from agrarian reform and training programmes in new agricultural methods because Western experts have assumed the existence of a pattern of responsibility for agriculture similar to that of their own societies, in which men are the main agricultural decision makers. This error has resulted in failure for many agricultural development projects. Even when included in development projects, women may be unable to obtain new technological inputs because local political and legislative attitudes make women less credit worthy than men. Where both men and women have equal access to modern methods and inputs there is no evidence that either sex is more efficient than the other. The introduction of high-yielding varieties of crops may increase the demand for female labour to weed and plant while leading to an increase in landlessness by widening the economic divisions between farmers. On the other hand, technological change in post-harvest processing may deprive

Table 4.2 The impact of agricultural modernization on women

Changes in the rural economy	Changes in women's socio-economic condition					
	Property ownership	Employment	Decision making	Status	Level of living and nutrition	Education
I Structural Capitalist penetration of traditional rural economy	Loss of rights of usufruct. Increase in landlessness. Sale of small properties.	Proletarianization of labour. Increase in male migration. Increase in employment of young unmarried women in agro-industries, urban domestic employment and multinational industries. Decline in job security. Increase in overall working hours. Triple workload of women as farmers, homemakers and wage labourers.	Increase because of male migration and economic independence of young women.	Increase in proportion of female heads of households because of male migration.	Increased dependence on remittances from migrants and employed children. Loss of usufructuary rights leads to decline in subsistence production and substitution of store-bought goods for home-produced.	Possible increase for daughters as their economic role becomes more important. May decline as increased burden on mothers forces daughters to take on more household tasks.
Land reform Colonization	Women generally not considered in redistribution of land. Loss of inheritance rights.	Decline in women farm operators. Increase in female unpaid family workers.	Decline because of patriarchal nature of colonization authorities.	Decline because of loss of economic independence.	Increased dependence on male head of household often leads to decline in family nutrition level despite possible increase in disposable income of family.	Increase if improved facilities, decrease if increased physical isolation. Children may be needed to work farm.
Socialist transformation of the rural economy	Theoretically no gender differentiation in rights on collective and state farms or to remaining privately owned land.	Increase in female participation in employment outside household. Increased specialization and division of labour. Women do most of work on family plot. Women's work on state farms undervalued. State support for communal childcare but women still have triple workload.	Separation of male and female decision-making units. Women underrepresented in state farm and collectives planning committees.	Increase because of 'de iure' equality between sexes and women's increased economic role.	Increase because of role of state in nutrition, health, housing, etc.	Women have equal access to all levels and types of education.

Table 4.2 continued

Changes in the rural economy	Property ownership	Employment	Changes in women's socio-economic condition		Level of living and nutrition	Education
			Decision making	Status		
II Technical						
Green revolution – new seeds and livestock breeds, pesticides, herbicides, irrigation	May lose usufruct rights as land is used more intensively. Land owned by women is often physically marginal and not suitable for optimum applications of new inputs.	Women exclude themselves from use of chemicals because of threat to their reproductive role. New crops may not need traditional labour inputs of women. Women generally displaced from the better-paid, permanent jobs.	Decline. Training in new methods in agriculture limited to men. Use of new technology and crops generally subsumed by men. Women farmers equally innovative when given opportunity.	Increase in family income may allow women to concentrate on reproductive activities. In patriarchal society this increases status of male head of household.	New crops may be less acceptable in family diet and nutritionally inferior because of chemicals.	Increase in additional disposable income of family may be used for children's education.
Mechanization	Women operate smaller farms in general and so may not find it economic to invest in new implements.	Women usually excluded from use of mechanical equipment. Women farmers have difficulty obtaining male labourers.	Decline.	Decline because of reduced role on farm and downgrading of female skills.	New implements not used for subsistence production.	Growth of interest in mechanical training but limited to males.
Commercialization of agriculture and changes in crop patterns	Female-operated farms tend to concentrate on subsistence crops and crops for local market. Tend to remain at small scale.	Decline because technical inputs substituted for female labour.	Decline because less involved in major crop production activity.	Decline.	Decline because cash crops take over land traditionally used for subsistence production by women. Male allocates more income to developing enterprise and for personal gratification than to family maintenance.	Increased time available for education.
Post-harvest technology	New equipment owned by men.	Women's traditional food processing skills no longer in demand. May employ young women in unskilled jobs in agro-industries.	Decline because ownership of equipment and skills passed to men.	Decline because female skills downgraded.	Decline because loss of women's independent income from food processing activities. New product may be nutritionally inferior. Women deprived of use of waste products for animal feed and so lose important part of traditional family diet.	Increased time available for education.

Table 4.2 continued

Changes in the rural economy	Property ownership	Employment	Changes in women's socio-economic condition		Level of living and nutrition	Education
			Decision making	Status		
III Institutional						
Credit institutions	Lack of access to credit limits expansion of female enterprises.	May allow increase in number of paid workers on farm and release of female family labour.	Decline because patriarchal control of credit.	Decline.	Increase if credit used wisely. May decline catastrophically if land used as collateral and lost to credit institution.	Increased demand for 'education as agricultural enterprise grows and horizons broaden.
Co-operatives	Women often excluded from membership of co-operatives.	Work on co-operative often undervalued.	Decline because not included in co-operative decision-making boards.	Decline.	May lead to general improvement in level of living.	Increased time available may increase demand for education from women.
Marketing and transport	Ownership of transport facilities generally male but if women maintain marketing role may invest in a vehicle. If own land women may retain marketing role.	Decline in traditional role in marketing with decline in production for local sale and increased size of market area. Difficult for women to travel long distances because of time demands of family and physical dangers.	Generally excluded from marketing decisions as community production is incorporated in national and international system.	Decline because of loss of traditional role.	Decline because loss of women's income from marketing. Exposure to imported, manufactured goods reduces income from craftwork such as pottery and weaving. Use of imported foods may reduce nutritional level but may also reduce time spent by women in food preparation.	Need for numeracy and literacy may lead to increased appreciation of advantages of education. Time released from household and marketing duties may be used for education.

women of a traditional income-earning task. Although the relationship is
not simple, new technology and crops seem in most places to benefit men
rather than women, especially when not accompanied by political change.

The growth of export production has provided new jobs for women.
Mexico produces early strawberries and tomatoes for the North American
market and women provide most of the labour used in picking, grading
and packaging these products. In Brazil, Mozambique and Sri Lanka
cashew nuts are produced for export and women process the crop. This
is an unpleasant job involving removing the nut from its protective outer
casing which contains an acid harmful to the skin. In northeast Brazil
women work in cashew processing factories with no protective clothing
supplied. Not surprisingly the workforce has a very high turnover rate.
The workers in Brazil are predominantly young, unmarried women but
in Sri Lanka they tend to be older, married women. In some villages in
Sri Lanka these women have banded together and obtained adequate
credit to enable them to set up as petty commodity producers of

Plate 4.5 Processing Brazil nuts is marginally less unpleasant than working with
cashew nuts but the job is still done by young women. Brazil nut workers in Rio
Branco, Acre, Western Brazil
Source: Mark Edwards/Still Pictures

cashews. In this way they eliminate the share of the profit which used to go to middlemen and have a home-based, income-earning occupation which can be integrated with household tasks and childcare. The additional income accruing to these women has given them a stronger voice in community affairs and changed the traditional patriarchal power balance in many households. Once again we have an example of the differential and unpredictable impact of change on women.

Gender differences in time budgets

Women on small farms in the Third World often have a triple burden of work. They are expected to carry out the social reproduction of the

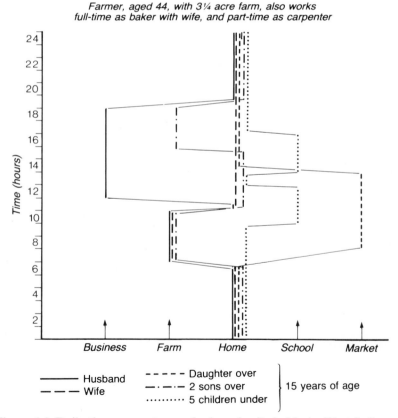

Figure 4.2 Daily time use patterns of a farm family in Nevis, West Indies

household which may include long treks to fetch water and firewood, wearisome journeys to take children to a distant clinic, time-consuming preparation of traditional food and the depressing job of trying to keep poor-quality housing clean. At the same time, rural women usually have to provide unpaid labour on the family farm and to earn money by working on another farm or by selling surplus produce.

Figure 4.2 is a time–geography diagram showing how daily tasks are distributed in time and space among family members on a 1.3 hectare farm on the Caribbean island of Nevis. The mother bakes bread for the village in a charcoal-fired oven built in her back yard. Her husband helps when he is not called away by his part-time job as a carpenter. The elder daughter is now old enough to take over her mother's former job of selling the excess produce from the family farm in the market in the nearby town. The older sons now do much of the farmwork, releasing their father for his income-earning off-farm job. Mother works on the farm a short distance from the house and as a baker but has to be at home to give her younger children lunch when they return from school in the middle of the day. She rarely has the time to move outside the immediate environs of her home and farm. It is possible to see from Figure 4.2 the effect of stage in life cycle and the number of dependants in a family on the mother's workload. In the example, the presence of adult children in the family has eased the burden on the parents of farmwork and marketing and the time–space demands of childcare are reduced as all the younger children are in school.

Time use studies reveal daily, weekly and seasonal fluctuations in the demand for labour and clarify the trade-off between productive work, household maintenance and leisure at different times of year and in various farming systems. It also makes it possible to identify age, sex and season-specific labour constraints which may need to be overcome if a new project is to be successful. In most rural communities women work longer hours than men and have less leisure time. In the Gambia, women spend 159 days per year in work on the farm while men spend only 103 days a year in farmwork and the women also spend an additional four hours per day on household maintenance and childcare. In Trinidad it was found that women worked longer hours than men on rice and dairy farms but less than men on sugar cane holdings.

When there is a labour shortage at busy times of the agricultural year women will often be expected to sacrifice their remaining leisure time for additional farmwork, acting as a reserve labour force. In Nevis, in the West Indies, for example, weekly hours worked by women on the farm equal 72 per cent of male hours in the busy season but only 66 per cent in the less busy period of the agricultural year. In Sri Lanka, on the other hand, women work longer hours than men in the peak season and only very slightly less in the slack season (Table 4.3). Despite their major contribution to agricultural production, women still do almost all the housework and the collecting of wood and water and are responsible for most social and religious duties. Consequently they have much less time than men for leisure throughout the year and at the peak agricultural season sacrifice an hour a day of their sleep and leisure time for extra farmwork despite reducing the time they spend on their reproductive tasks. Thus women's workhours in rural areas of the dry zone of Sri Lanka average over 18 hours per day in the peak season compared to 14 for men.

Table 4.3 Gender divisions of time use in the dry zone of Sri Lanka

| | Peak season | | Slack season | |
| | Male | Female | Male | Female |
		hours per month		
Agricultural production	298	299	245	235
Household tasks	90	199	60	220
Fetching water and firewood	30	50	30	60
Social and religious duties	8	12	15	15
Total workhours	426	560	350	530
Leisure/sleep	294	160	370	190

Source: A. Wickramasinghe 'Women's role in rural households and agriculture in Sri Lanka' in J. H. Momsen (ed.) *Geographical Studies in Women and Development*, (provisional title) to be published by Routledge

In Zambia, today, women do 60 per cent of the agricultural work but domestic chores take twice as long as farmwork. Food preparation is one of the most onerous tasks and development had little beneficial effect on the time and energy needed for domestic labour. The heavy burden of both productive and reproductive work has contributed to a high incidence of ill-health among rural Zambian women and to poor family nutrition. Over 50 years ago, during a period of high male migration, it was noticed that at times of heavy demand for agricultural labour women frequently failed to cook meals and the family went hungry, despite plentiful supplies of food, as the women were too exhausted to collect firewood and water and to gather the relish needed for the meal. Contemporary research suggests that the recent

Plate 4.6 Cooking in Burkina Faso
Source: Elsbeth Robson, University of Oxford

intensification of women's labour input on farms as a consequence of the introduction of maize cash-cropping may have contributed to the apparent increase in child malnutrition because women have less time to prepare meals, especially weaning foods, for their families.

Lack of time is resulting in a de-intensification of farming in central Java. Increased need for regular and reliable cash income is pushing both men and women farmers into paid non-farm employment and poor women farmers in particular find that the combined demands of reproduction and non-agricultural production leave little time for farmwork.

Plate 4.7 Cooking in the Ivory Coast
Source: Elsbeth Robson, University of Oxford

Case study D

A woman tobacco farmer in Trinidad

A full-time tobacco farmer, Mrs T., a middle-aged widow, lives on land leased from the tobacco company, a multinational firm. This firm provides seed stocks, fertilizers and pesticides and dictates the number of times land should be ploughed and when crops should be sprayed, fertilized and irrigated. Since the company is the sole purchaser of cured tobacco in Trinidad it can establish grades and refuse to purchase an unsatisfactory crop. Farmers who disobey the company's orders are quickly removed from their leased farms. Tobacco demands high labour inputs and provides a poor standard of living. Mrs T., like most tobacco farmers, lives in a single-storey, dirt-floored wooden home without electricity or running water. She depends on income from the sale of vegetables

Case study D *(continued)*

to support her family and to pay the high cost of services and inputs supplied by the tobacco company.

She was born to Trinidadian sugar cane farmers of East Indian origin, in one of the major rice and cane areas. She had barely three years of schooling and before the age of 10 she began doing housework and later helped in the cane fields, 'breaking banks' and cutting cane. Her marriage to a 16-year-old labourer was arranged when she was only 13. At her husband's family home she assumed many domestic tasks and was also forced to work in the cane fields. All her earnings went to her mother-in-law. Later, she was hired by a tobacco farmer and learned tobacco cultivation. There are eight children still living at home as well as one daughter-in-law, who is expected to assume the same tasks Mrs T. performed as a young bride. For additional income, Mrs T. grows vegetables all year and sells them daily at the local market.

Her typical working day begins at 3.00 am when she starts to clean the house, prepare and serve the family breakfast and get the children ready for school. At 7.00 am she leaves the house to work in the tobacco or vegetable fields until noon. Between noon and 1.00 pm she eats lunch and rests. From 1.00 to 3.00 pm she sells in the markets and from 3.00 to 6.00 she prepares produce for sale the next day. Most of the housework is done by her elder daughters and the whole family shares work during the tobacco harvest.

Mrs T.'s one ambition is to be able to purchase a piece of land for her family and to educate her children but it appears unlikely that she will succeed.

Source: Indra S. Harry (1980) 'Women in agriculture in Trinidad', unpublished M.Sc. thesis, University of Calgary, Canada

Women in the plantation sector

Plantations are the organizations which enabled the integration of many parts of the Third World into the periphery of the world economy. The archetypal example is the sugar plantation in the Caribbean and Brazil in the seventeenth century, but plantations were also established at a

Plate 4.8 Women in St Lucia, West Indies loading
bananas for export to Britain
Source: The author

later date, for rubber in Malaysia and tea in Assam and Sri Lanka.
Female labour has been very important in this sector. Under slavery,
women came to form an ever greater proportion of the field labour force
on Caribbean sugar plantations and at emancipation were in a majority.
Even today plantation labour remains an important source of income
for poor women in rural areas of the region. In Malaysia women account
for over half the plantation labour force and this proportion is increasing. The importance of women as plantation labourers today reflects a

decline in available male labour and the lack of alternative employment opportunities for rural women.

Work on plantations is generally unskilled, poorly paid and seasonal. In many cases, the only way for the household to survive is for women and children, as well as men, to work as a unit to maximize output. This was the traditional method on Brazilian coffee plantations for harvesting the beans and is used on Malaysian rubber plantations. In both cases the estate provides a small subsistence plot for the family and expects the women and children to act as a reserve pool of labour which can be called on at peak periods of demand. Such an employment pattern means that women are encouraged to have many children and cannot afford to allow these children to take time away from plantation labour to attend school.

Case study E

Women plantation workers in Sri Lanka

Tamil workers from India were brought to Sri Lanka by the British in the nineteenth century to work on the newly developed tea, rubber and coconut plantations. Today they account for 5.6 per cent of the total population of the country but three-quarters of them live on plantations and they are spatially concentrated on the tea plantations of the central part of Sri Lanka. Among the Tamil plantation workers the female participation rate in the workforce is 54 per cent compared to only 17 per cent for rural non-estate women. However, the physical quality of life of these women workers is much inferior to the national average. The Tamil plantation workers have above average maternal and infant mortality rates, very low fertility rates, high female illiteracy rates and are the only group in Sri Lanka in which women have a lower life expectancy than men. The group's general poverty has been exacerbated for women by the cultural partriarchal norms which have resulted in women's subordination and reduced their access to basic needs.

The female tea plucker's day begins before sunrise. She gets up around 4.00 am to prepare breakfast and lunch, clean the house, and get the children ready for creche and/or school. The morning meal consists of homemade bread (*roti*) with a watery curry and

Case study E *(continued)*

tea. By 7.00 am the tea pluckers are at work in groups and keep filling their baskets with leaves until the tea break from 9.30 to 10.00 am. The lactating mothers visit the creche to nurse their babies and then resume work until 12.30 or 1.00 pm. The woman worker takes the load to the weighing shed, visits the creche, nurses her baby and goes home for the midday meal which she has prepared the night before. She returns to the field by 2.00 pm and continues to pluck leaves until 4.30 pm. She takes the load to the weighing shed and waits her turn. She visits the creche to collect the children and returns home around 5.30 pm. She then starts the evening chores: cleans the house, prepares the evening meal and the next day's midday meal, feeds the children, cleans them and washes the clothes and puts the children to bed. She is often the last to go to bed around 10.00 or 10.30 pm. She sleeps on a sack on the floor as there is usually only one cot in the one-roomed house, which is used by her husband.

The per capita calorie intake of plantation workers is one of the highest in Sri Lanka but 60 per cent of the workers suffer from chronic malnutrition, a rate which is almost twice as high as in the rest of the population. Part of the explanation for such apparent dietary deficiencies seems to lie in the subordinate position of women in the family, their heavy workload and their lack of time for food preparation. The woman tea picker's work involves climbing steep slopes carrying a weight of up to 25 kilos in a basket on her back and constant exposure to rain, hot sun and chill winds. She has neither time nor energy to prepare nutritious meals for her family. Nor does she have time to purchase a variety of foodstuffs as the plantation store is expensive and has few items in stock while other shops are too far away to be visited frequently. As meat and fish cannot be stored because of lack of refrigeration these items are eaten rarely. Until 1984 female workers, although working longer hours than male plantation workers, received lower wages. They now earn equal daily wages but women rarely have time to queue to collect their wages and so their husbands collect them for them. In many cases the women complain that their husbands do not give them their wages but spend the money

Case study E *(continued)*

on alcohol and gambling. Even maternity payments are collected by husbands and it is doubtful if these are spent on supplementary food for the lactating mother or new-born baby.

Plantation workers have strong unions but although more than half the workers are female there are no women union leaders. Better access to basic needs such as improved schooling, widespread health care and more sanitary living conditions seem to receive low priority in union demands. Household technology remains primitive and time consuming and the allocation of time between increased wage work and household chores is a constant balancing act for women plantation workers.

Source: Vidyamali Samarasinghe 'Access of female plantation workers of Sri Lanka to Basic Needs Provision' in J. H. Momsen (ed.) *Geographical Studies in Women and Development*, (provisional title), to be published by Routledge

Women as rural traders

Not only do women produce and process agricultural products but they are responsible for much of the trade in these, and other goods, in many parts of the Third World. As is shown by Figure 4.2 the sale in the local market of surplus production is an important element in small-scale agriculture. In the Caribbean this role developed under slavery and was seen as so important to the local economy that the plantocracy allowed women traders the freedom, despite their slave status, to travel around within each island. Thus women had a powerful position as carriers of news and information between slave plantations. In many parts of the world, women continue to play an important role as rural information sources and providers of food to urban areas. This may involve food from the sea as well as from the land. Although women rarely work as fisherpeople they are often involved in net-making and the preparation and sale of the catch. Thus, in many cases, it is women, especially in West Africa, who through the trading of goods, act as a major link and source of information between rural and urban sectors of Third World economies.

Key ideas

1 There are marked regional patterns of female participation in agriculture, ranging from a high 87 per cent in low-income African countries to 14 per cent in Latin America.
2 The extent of women's involvement in subsistence production is a function of the nature of local farming systems, access to resources and the degree of overall commercialization of the agrarian sector.
3 All farming systems have gender-specific agricultural tasks.
4 Women work longer hours than men in rural areas because they have a triple burden of household maintenance, farmwork and paid labour.
5 Agricultural development is not usually beneficial to women. They are often ignored by change agents and may lose access to their traditional resources.
6 Technological change tends to reduce the role of women in agriculture. Tasks that are mechanized become male jobs. Labour and time-saving technology has differential effects on women of different classes.
7 Increased workloads and time pressures for rural women have led to health problems, the de-intensification of agriculture and worsening family nutrition in many parts of the Third World.

5
Women and work in urban areas

Throughout the world more and more people are living in urban areas and many of the world's largest and fastest growing cities, such as Mexico City and São Paulo, are in Third World countries. This trend towards urban residence has been accompanied by an increased visibility and variability of jobs done by women.

Before looking at the interrelationships between development and changes in women's work in urban areas we must consider two major deficiencies in the data. First, most national and international employment statistics are collected on the basis of economic sectors rather than job location so we must make the crude assumption that all non-agricultural employment is in urban areas. Secondly, these statistics must be treated with caution because a significant proportion of non-agricultural production, sales and service jobs done by women takes place outside a formal workplace and may not be officially recorded. For all jobs, the bias is generally towards an under-reporting of women's work. This under-reporting tends to be greatest for self-employed women involved in craftwork or street vending, and for occupations on the edge of legality such as prostitution, while wage or salary employment for a major employer tends to be most accurately recorded.

In Third World countries, the proportion of the labour force in non-agricultural jobs is estimated to have increased from 27.4 per cent of the labour force in 1960 to 40.9 per cent in 1980. The proportion of these jobs going to women was highest in Latin America (35.3 per cent), and lowest in Arab countries (11.7 per cent). The averages for Africa south of the Sahara, and Asia at 24.3 per cent and 27.9 per cent respectively, fell between these extremes. There was great variation from country to country with Haiti recording 66.4 per cent and Singapore 44.9 per cent while, at the opposite end of the scale, Libya had a figure of 5.4 per cent and Bangladesh 5.5 per cent.

There is no clear relationship between higher levels of development, urbanization and increased female employment. Indeed, in India and parts of Africa urban growth is associated with a decline in overall female labour force participation. Underrepresentation of women among non-agricultural employees tends to be greatest in the least developed countries, indicating a time-lag effect on the employment of women in modern occupations.

Female marginalization

It is often suggested that women's role in production becomes progressively less central and important during capitalist industrialization in developing countries. There are four dimensions of 'marginalization' as it is applied to urban female employment. Firstly women are prevented from entering certain types of employment, usually on the grounds of physical weakness, moral danger, or lack of facilities for women workers. Secondly, marginalization can be seen as 'concentration on the periphery of the labour market' where women's employment is predominantly in the informal sector and the lowest-paid, most insecure jobs. Thirdly, workers in particular jobs may become so overwhelmingly female that the jobs themselves become feminized and so of low status. A fourth dimension is marginalization as 'economic inequality'. This aspect refers to the economic distinctions which accompany occupational differentiation, such as low wages, poor working conditions and lack of both fringe benefits and job security in work thought of as 'women's'.

Clearly the concept of marginalization is complex and is often difficult to identify empirically. It is also a relative concept varying from place to place and cannot be used to predict changes over time. However, it remains useful as a descriptive tool. Female marginalization is usually blamed on efforts by capitalists to minimize labour costs, but historical, cultural and ideological factors are also important.

Case study F

The effect of economic growth on the labour market position of women – Lima, Peru 1940–72

The period from 1940 to 1972 was one in which modern industry became established and educational opportunities for women expanded in Lima, the capital city of Peru. The proportion of women with primary and/or secondary education increased from 77.4 per cent in 1940 to 82.2 per cent in 1972 while the proportion of men at this educational level decreased from 86.5 per cent to 84.4 per cent. Theoretically by 1972 women should have suffered only a very minor disadvantage when competing with men for the vast majority of jobs which require only a primary or secondary education. However, labour force participation rates were highest among the few women with post-secondary education at 53.6 per cent followed by illiterate women at 35.5 per cent, while those with secondary education had the lowest participation rate of 26.8 per cent. This suggests that the labour market for men and women was divided and that barriers for women were not based on skill levels.

Family structure and a civil code which relegates women to the domestic sphere and requires wives to obtain the permission of their husbands to work outside the home continues to constrain women's role in the labour market. This familial restriction has least effect on poor, usually illiterate women whose earnings are essential for family survival and on women with professional training. Industrialization and education failed to change family structure or fertility rates during this period. Women's share of the unskilled workforce in manufacturing fell while they increased their share of professional and service sector jobs. Employment in domestic service became almost entirely female. The symbiotic relationship between these two groups of women strengthened, with the professional women's dependence on a continuous supply of cheap female domestic labour producing an all-women supply and demand market.

Source: A. MacEwen Scott (1986) 'Economic development and urban women's work: the case of Lima, Peru' in R. Anker and C. Hein (eds.) *Sex Inequalities in Urban Employment in the Third World*, London: Macmillan

Gender divisions of labour

The main explanations for the disadvantaged position of women in urban labour markets fall into three groups based on different theoretical viewpoints: those of neo-classical economics, labour market segmentation and feminism.

Neo-classical economic theory

This assumes that in competitive conditions, workers are paid according to their productivity. It follows from this assumption that observed male–female differentials in earnings are due either to the lower productivity of women or to market imperfections. This approach also assumes that women have lower levels of education, training and on-the-job experience than men because families tend to allocate household resources to the education of male family members while expecting the females, as they grow up, to spend their time on housework and childcare for which training is not required. So neo-classical theory explains gender differences in employment in terms of differences in human capital where women are disadvantaged because of their family responsibilities, physical strength, education, training, hours of work, absenteeism and turnover. However, it has been shown empirically that these variables can explain only a part of the wage gap between men and women.

The neo-classical approach has been criticized on the basis of three of its underlying assumptions. Firstly it assumes that the gender-based wage differential can be largely overcome by improving the education and training of women. Where differences in education levels are very marked, as in predominantly Muslim countries, this may have some initial effect. However, in the long run the result may be to raise the level of education in 'women's jobs' rather than to decrease pay differentials. A second implicit assumption is that men and women have equal access to the labour market and compete on equal terms for job opportunities. This ignores the gender-based segregation of the labour market which exists in all countries and does not appear to decline as gender differences in education levels even out. A third underlying assumption is that women's labour force participation is of necessity intermittent because of their 'natural' childbearing role. Yet only pregnancy and breastfeeding are biologically restricted to women and in Third World countries many women are able either to share childcare with relatives or friends, or to employ domestic servants, or to keep children with them while they work, or they have access to a free creche.

Theories of labour market segmentation

This approach emphasizes the structure of the labour market in explaining sex inequalities in employment. It assumes that the labour market is segmented by institutional barriers but within each segment neo-classical principles still apply.

One such division of the labour market is into primary and secondary sectors. Primary jobs are those with relatively good prospects of promotion, on-the-job training and pay while secondary sector jobs are poorly paid and have little security. Because of the perceived higher turnover of women they are more likely to be recruited into secondary sector employment while men will be sought for primary sector jobs. Yet turnover and absenteeism are higher in low-level, boring, dead-end jobs, such as those of the secondary sector, where women are concentrated, and so these aspects of employee reliability may be explained by sex differences in type of occupation rather than by inherent characteristics of women.

Other factors influencing this gender segregation include the better organization of male workers to defend their skills and income differentials, their resistance to competition from cheaper (often female) labour and the role gender relations and patriarchal ideologies play in the control structure of the firm.

In many parts of the Third World this differentiation within the capitalist sector is given less emphasis since women tend to be generally excluded from employment in this sector. The industrialization process in the Third World is capital intensive and is dominated by foreign capital and imported technology. This type of industry, often referred to as the 'modern' or 'formal' sector, has a low level of labour absorption and is biased against the employment of women because of their lack of formal educational qualifications, their supposed lower job commitment and because capital-intensive skills tend to be considered 'male' skills. Female employment is concentrated outside this sector in the 'informal' or 'traditional' part of the labour market. The production arrangements in this sector include self-employment, out-working, family enterprise and household service which offer the flexibility needed by women in combining the demands of their reproductive and productive activities on their time. It also provides flexibility of labour supply for large-scale manufacturers who can subcontract work out to small-scale enterprises at times of peak demand.

Plate 5.1 Building labourers, India
Source: Janet Townsend, University of Durham

However, this model ignores the wide range of technologies which exist in modern industry, some of which, such as light assembly work, discriminate in favour of women. It also ignores the increased demand for women workers created by the expansion of the modern sector in the female-dominated clerical, teaching and nursing occupations. It does not explain the high degree of gender segregation within the informal sector nor the frequent movement of individuals between the formal and informal sectors.

A segmentation of the labour market based on gender may also be observed. The existence of two separate labour markets for men and women tends to restrict women's occupational choices. To the extent that there is an oversupply of candidates for women's jobs this may maintain lower pay levels in this segment of the labour market while restricting competiton within the male-dominated segment and thus keeping wage rates relatively high for men. The sex of the workers may of itself lead to women's jobs being defined as unskilled while jobs filled by men are defined as skilled.

These economic theories tend to assume that gender roles in society are fixed and are the basis of women's disadvantaged position in the labour market. This can lead to the circular argument that because women are not able to earn as much as men in the workforce it makes economic sense for them to stay at home doing unpaid domestic labour. It is clear that economic theories cannot fully explain gender differences in the labour market and much of the marginalization of women is the result of discrimination based on societal prejudices.

Feminist theories

Feminist theories emphasize the importance of social and cultural factors in restricting women's access to the labour market. These approaches tend to see the interaction between the reproductive and productive roles of women as a key issue rather than a fixed condition. The allocation of housework and childcare to women persists in most societies even though women's participation in the labour market is increasing. Female labour force participation in urban areas affects household composition: families tend to be smaller and there may be a shift away from nuclear families to both extended families and female-headed families. At the same time, domestic help is becoming scarcer and more expensive and educational opportunities, especially for daughters, are increasing and so children have less time to help their mothers in the home. Consequently, the burden of domestic responsibilities falls ever more heavily on one particular woman in the family. Few poor Third World homes have the domestic appliances commonly available in industrialized countries and housework is a very heavy burden. It is estimated that married women in Malaysia who do housework and are in paid employment outside the home spend, on average, 112 hours per week working, while the equivalent figure for the United States is 59 hours. Thus women's handicap in the labour market because of domestic responsibilities may be growing rather than diminishing in many Third World cities.

Sexual harassment may be an even greater problem in the Third World than in developed countries. In traditional societies, a woman who moves out of her accepted family role in order to take a job may be seen as a 'loose' woman. Men who are not used to meeting women in a work situation may fall back on gender-based social expectations and treat their workplace female colleagues as sexually available. Men in supervisory positions may demand sexual favours in return for job security and this may contribute to high turnover rates for women workers. Those women most in need of paid employment may be

victimized by sexual harassment, as the option of resignation from the
job, which may be their only means of escape, is often not open to them.
The ghettoization of women into certain sectors of the economy may be
encouraged by fathers who want to protect their young daughters'
reputations while at the same time needing to send them out to work to
contribute to the family income.

Non-economic exclusionary measures are sometimes political and
legal but most often are based on familial ideology and are sanctioned
by informal controls such as gossip or ridicule. Employment of women
in occupations such as teaching and nursing is seen as an extension of
their domestic role and so tends to be devalued. In many jobs qualities
attributed to men, such as physical strength, are valued more highly
than those characteristics thought of as female, such as manual dexterity
and docility.

Barriers to women's participation in the urban modern sector

Clearly, certain aspects of social, economic and cultural norms determine women's ability to participate in urban employment in Third
World countries. Modern industry is spatially separated from the home
and involves a standard fixed pattern of working hours. Both characteristics cause problems for women with children. In industrialized
societies, in recent years, women with family responsibilities have
sought a solution in part-time work but this is generally discouraged by
employers in the developing world. Furthermore, daily working hours
are often longer and paid holidays shorter in the Third World. Thus
many women put together multiple self-employed occupations in order
to gain an adequate income or seek work in the informal sector because
of its flexibility. However, the relatively high participation rate in the
modern sector of women with post-secondary schooling indicates that
women with well-paid jobs are able to cope with the demands of such
work because of the availability of cheap domestic help.

The burden of domestic work bears most heavily on poor women.
They are usually forced to depend on their own mothers, female friends
or older children for assistance with childcare. A few countries do
attempt to provide workplace or government funded creches. Only
Cuba appears to be ideologically committed in its Family Code to
reducing women's double burden of productive and reproductive work
by expecting husbands to undertake a fair share of household chores,
but this legislation has yet to become fully effective.

Protective legislation applicable to modern industry, such as work-hours and maternity leave, may limit women's work opportunities by raising the cost of female labour. Women are underrepresented in trade unions and they do not generally hold positions of office, so it is not surprising that issues concerning women are rarely taken up by trade unions.

There is considerable evidence of employer discrimination against women. Sometimes this is justified by the employer on the grounds of perceived lower productivity and higher absenteeism and turnover of women. The evidence to support these perceptions is not clear and varies from place to place. Where problems can be noted they are generally related to the family responsibilities of women.

Many employers have a preconceived idea of types of jobs suitable for women. They consider only a very narrow range of jobs as open to women and in this way women's opportunities are more restricted than those of men. Lack of physical strength and inability to supervise are the main reasons given for restricting jobs for women. The advantages women offer are seen as related to a willingness to work for lower wages than men and their greater docility. Employers also think women are most suited to jobs related to household skills or where femininity is an advantage, as in the case of waitresses. These stereotyped views limit employment opportunities for women.

Differences in education level also hinder women from entry into the best-paid jobs. However, this may be a self-fulfilling situation for where it is perceived by parents that the best jobs go only to educated males it may be thought that investment in a daughter's education is a waste of money. Lower levels of education among women do not explain all the difference in male and female earnings and it may be concluded that equality of education is a necessary but not a sufficient condition for equality of pay.

The assumption by policy makers that men are the main providers for the family means that where there is high unemployment, jobs will be found for men before women. When there is a recession women are usually the first to lose their jobs. Most Third World countries have higher unemployment rates for women than for men. Yet about one-third of households are headed by women who often have to support themselves and their children. Single women workers have often been found to contribute more than their brothers to the income of their family. The continuation of the myth that men are able to be the sole breadwinners perpetuates the secondary status of women in the labour market.

Case study G

Colombia – help for working mothers

When it rains in Bogotá, the paths that wind through the slum areas are awash with mud and litter, the tin-roofed shacks leak and people huddle inside their homes, waiting for the storm to pass.

Many of them are women struggling alone to bring up their children – widowed, divorced or separated in the mêlée of urban life. Even a woman with a husband cannot always rely on him for support when so many men are unemployed: without one it is even harder. Few have the skills to find a salaried job or set themselves up in business. Many are as trapped in their homes by their domestic responsibilities as they are by the teeming rain that beats on the tin roofs above their heads. There are babies and young children to look after, older relatives to tend, and cooking, cleaning, washing to be done.

A series of pilot projects in four of the poorest slum areas of Colombia – Bogotá, Cali, Medellín, and Barranquilla – aims to give around 10,000 of these women an escape route. The programme encourages women to join one of over 30 groups giving credit and training in the management of small businesses. Four community centres provide creches, care for old people, laundries and cafeterias. These facilities are run by women from some of the groups as just a few of a variety of new income-generating activities sparked by the programme. The community centres also provide a focus for health, family planning and nutrition education activities and a place where women can obtain health services and supplies of contraceptives.

Source: United Nations Fund for Population Activities (1986) *Women's Projects Funded by UNFPA in 1985*, New York: UNFPA

Women in cities have to cope with the spatial separation of home and work often without the support networks of relatives which exist in rural areas. The double burden of production and reproduction has led to both female interdependency as mistress and servant, and to the growth of female support groups to share the burden of family responsibilities.

Key ideas

1 The female marginalization thesis is empirically untestable but is useful as a descriptive concept.
2 Sexual stereotypes limit women to a narrower range of jobs than is available to men.
3 Equality of education is a necessary but not a sufficient condition for equality of pay.
4 The spatial separation of home and work in cities acts as a constraint on women's employment opportunities.
5 Discrimination and the belief that women are only secondary income earners results in women workers having higher unemployment rates and lower wages than men.
6 Female economic activity affects household composition.

6
Spatial patterns of women's economic activity

Officially recorded female economic activity rates are usually higher in cities than in rural areas, especially in Latin America. More women are moving into paid employment as traditional restrictions break down, demand for new consumer goods spreads and economic pressure on families worsens. Figure 6.1 shows the global pattern of women's participation in the labour force. Clearly the greatest variation occurs in the Third World.

It has been suggested that the relationship between development and female employment follows a 'U'-shaped curve with economic activity of women being highest in both least developed and post-industrial societies, while it is lowest in those countries at a middle level of development as women move out of agriculture. The map reinforces this model showing that Scandinavia and sub-Saharan Africa have similar proportions of women in the labour force although at opposite ends of the development spectrum. However, at intermediate points cultural, political and historical factors intervene to reduce the applicability of the model.

Socialist countries such as Cuba and China have high participation rates for women. Africa has the greatest variation with very low rates in Muslim North Africa and very high rates in parts of Africa south of the Sahara. Latin America and South Asia have quite low rates while Southeast Asia has high participation rates. The high rates in Africa are associated with female farming systems but elsewhere it is urban jobs which dominate. In Latin America women are employed mostly in the

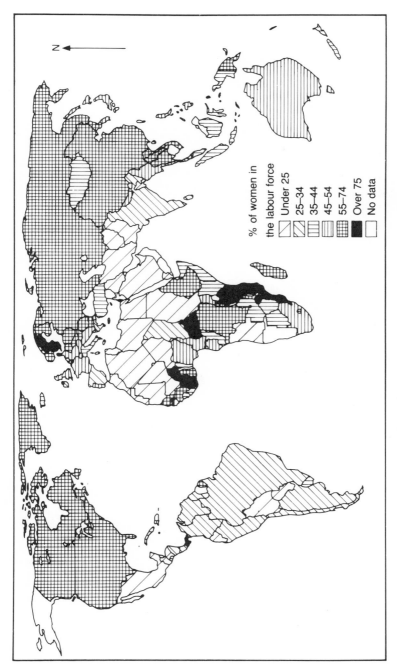

Figure 6.1 Women's participation in the labour force

% of women in
the labour force

Under 25

25–34

35–44

45–54

55–74

Over 75

No data

Plate 6.1 Rural schoolteacher outside schoolhouse by the Amazon River in Para, Brazil. The school has 20 pupils, most of whom travel to school by canoe
Source: The author

service sector especially in domestic service, teaching and clerical occupations. In Southeast Asia, the growth of world market factories employing mostly young women has led to an increase in female activity rates.

Age also affects the gender division of labour. In most societies male control of women's use of space is greatest during their reproductive years thus limiting their access to the labour market. Figure 6.2 shows

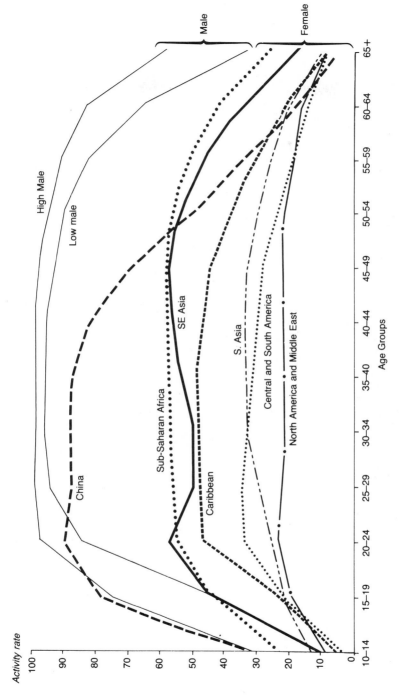

Figure 6.2 Female and male economic activity rates, 1980

that although male economic activity rates vary little from region to region, female rates show distinctive spatial patterns. In China, although the female activity rate is only slightly less than that of men, the decline with age starts earlier than elsewhere at about the age of 35. Consequently Chinese women have the highest female economic activity rate in the Third World between the ages of 10 and 50, but by the age of 65 the rate is lower than in any other region. In the Chinese countryside women's labour is so important that female slavery of both women and children is widespread. It is reported that the recent marked increase in the cost of a traditional Chinese wedding has led to single men in early middle age finding it cheaper to buy a bride or housekeeper.

In most parts of the developing world women reach their maximum level of economic activity in their early twenties while the maximum for men occurs a little later. In Southeast Asia there is a marked dip in the level of female employment between the ages of 20 and 45 which is the period during which women experience the most intensive childbirth and childrearing time demands. Other world regions do not demonstrate this so clearly and its importance in Southeast Asia may be a function of the high level of employment of young, single women in world market factories there.

In South Asia the maximum economic activity rate for women comes much later than elsewhere at age 45–9 and retirement also comes later, reflecting the early age of marriage and childbearing in this region. Early retirement from paid employment for women may occur because they no longer need the income as their children are grown-up and can support their parents. Or women may leave the labour force in order to take over the care of grandchildren and so release their own daughters for work outside the home. In any case these older women do not generally retire into idleness but take on new childcare and household responsibilities and often increase their labour input to the family farm.

The expansion of educational opportunities for women in recent years is reflected in the type of employment undertaken by women of different ages. Especially in Latin America and the Caribbean, where there has been a marked increase in women's access to education, younger women have moved into white collar, urban jobs which offer regular employment, pensions and status. Their mothers generally continue to work intermittently in unskilled work such as agricultural labour or trading. The greater financial independence of young women enables them to be less dependent on men and also less likely to see having children predominantly in terms of ensuring a future financial resource.

Plate 6.2 Woman in the market in St Lucia, West
Indies, selling handmade charcoal-fuelled pots used
for cooking
Source: The author

Industry

In most developing countries women have been moving out of agriculture and into industry faster than men. As a result the proportion of
women in Third World industry has risen from 21 per cent in 1960 to
26.5 per cent in 1980. In Hong Kong, South Korea, Taiwan, the
Philippines, Singapore, Thailand, Tunisia and Haiti the share of women
in the manufacturing labour force is more than 40 per cent while in no

Plate 6.3 Young women factory workers making electric rice cookers for export near Guangzhou (Canton), Southern China
Source: The author

industrialized market economy do women account for more than 31 per cent.

The influence of the international economy, as articulated by transnational manufacturing companies, has created a new market for female labour. Manufactured exports from developing countries have been dominated by the kinds of goods normally produced by women workers. Industrialization in the postwar period has been as much female-led as export-led. However, the international economy has put a premium on low wages so the benefits to women of increased employment opportunities are equivocal. Some 4 million young women are employed in light export-oriented industries mainly in South and Southeast Asia and Latin America.

Women workers are concentrated in light industries producing consumer goods ranging from food processing, textiles and garments to chemicals, rubber, plastics and electronics. In Egypt, Hong Kong, India, Kenya, the Philippines and South Korea over three-quarters of the female industrial labour force is employed in these seven industries.

Case study H

Attitudes to women factory workers in Malaysia

The share of manufacturing in the Malaysian Gross Domestic Product grew from 9 per cent in 1960 to 21 per cent in 1980 but perhaps its most spectacular aspect was the massive and sudden involvement of young, single Muslim Malay women from rural areas. Between 1957 and 1976 the number of Malay women in the manufacturing sector increased twelve times while the proportion of male workers declined.

Never before had Malay women left their traditional village occupations in such numbers. Most of these women came from families twice as large as the national average but with very low incomes. Three-quarters of the women chose to migrate to work in factories in order to reduce economic dependency on their households. Although factory wages were as low as those paid for agricultural work, they were more stable and offered fringe benefits such as subsidized meals, medical services, transport to work, uniforms, sports facilities and other leisure activities. A study showed that 56 per cent of the women migrated from their villages because they wanted to obtain a job and improve their standard of living while a further 19 per cent did so in order to gain personal freedom and independence.

For the manufacturers these employees have many attractions. They are aged between 16 and 20 and so are more easily disciplined than older women. They are single and are thought to be more dependable than married women and more available for overtime assignments. They are poorly educated but not illiterate and the traditional rural compliance to male authority makes them the naive, obedient and malleable workers the firms want.

They are paid only 69 per cent of male wages for the same job. They work 50 per cent more hours than women doing similar work in the West and receive only 10–12 per cent of the pay of Western workers. Low incomes lead to poor living conditions with overcrowding and few amenities. The combination of Western attitudes inculcated through factory work and living away from the protection of their family has led to involvement in social activities which are in conflict with traditional Malay Muslim values. Many parents

Case study H *(continued)*

are beginning to feel ashamed that their daughters are employed in an occupation that is rapidly acquiring a low moral and social status in Malaysian society.

In villages close to urban areas women can commute daily to factories and in these settlements both family and community conflicts arise. In the community studied, 41 per cent had perceived in factory workers negative personal changes, such as indecent dressing, liberality in social mixing, decreased standards of morality, devaluation of domestic roles and loss of interest in local affairs. On the other hand, 12 per cent thought that the factory workers gained by greater social exposure and being able to be self-supporting. Many people thought that Malay women should be encouraged to work in factories but 70 per cent of these people did not wish women of their own family to do so. Overall the source community accepted the utility of factory work in the short-term as an answer to immediate economic problems but rejected it in the long-term because of the social and moral disutilities that were developing.

Source: Amriah Buang 'Development and factory women – negative perceptions from a Malaysian source area' in J. H. Momsen (ed.) *Geographical Studies in Women and Development*, (provisional title) to be published by Routledge

Women also work in manufacturing outside the formal economy of the factories. Studies in Mexico City have shown that the number of women working in their homes producing items on contract for factories has increased during this decade. Women are employed to do simple, unskilled, labour-intensive tasks of assembly or finishing, requiring minimum use of capital or production tools. Working in the home allows women to carry out their productive and reproductive chores in the same location. The advantages of out-working for employers are the flexibility it gives them to respond to changes in demand and the reduction in labour costs. This work is on the edge of legality because of the absence of regulation which enables employers to pay below minimum wage rates and to avoid providing fringe benefits and workplace facilities required by law. The work offers no security but may be the only option for women trapped in the home with young children.

Women also work as petty commodity producers in both rural and urban areas. Like out-working, this work offers women flexibility of time and space as it allows them to remain within the home and can be combined with domestic chores. In traditional societies it may be more acceptable for women than employment which is located outside the home. Goods produced vary from textiles and garments involving weaving, lace-making, sewing and embroidery, to ceramic and food items. It builds on women's traditional skills and has been expanding recently as aid organizations offer assistance in the form of credit, training, design and marketing.

Women in the service sector

Underestimation of women's work is particularly severe in this sector. Even so official statistics indicate that women make up 27 per cent of the service workers for the Third World as a whole and 39 per cent in Latin America where 70 per cent of all economically active women are service industry workers. In general women work in health, education, catering, tourism and commerce at the lowest and worst-paid levels.

Informal sector work

Providing services in the informal sector involves many women. Work in this sector as traders, servants or prostitutes is often the only urban employment open to young, uneducated women from rural areas. There are distinct regional patterns in the dominant type of employment.

Women are especially important in retail trade in Africa and the Caribbean. Women make up 93 per cent of market traders in Accra (Ghana), 87 per cent in Lagos (Nigeria), 60 per cent in Dakar (Senegal) and 77 per cent in Haiti. Status as a market trader comes with maturity and mothers often pass on to daughters their bargaining skills and the goodwill of their customers. Although most traders in West Africa are women, they tend to concentrate on the sale of small quantities of home produced items in local markets while men control wholesaling and the long-distance trade in manufactured goods. Similar distinctive gender roles can be seen in the Caribbean where beach vendors selling to tourists fall into two groups: young men who sell jewellery or suntan oil and work as vendors for a few years because they enjoy the beach and

Plate 6.4 Market in Roseau, Dominica
Source: The author

the chance to meet young female tourists; and older women who braid hair or sell homemade clothing on the beach because the job does not have fixed workhours and can be combined with childcare.

In Latin America domestic service occupies many women. It is seen as an entry point into urban employment for female rural immigrants. Their lowly occupational status may be reinforced by racial discrimination. In Andean and Central American countries, servants are often Indian and may speak little Spanish in contrast to the Hispanicized families for whom they work. Domestic service is distinctive in that

Plate 6.5 Domestic service. Maid in Abidjan, Ivory Coast
Source: The author

women are both employer and employee. Employers prefer young rural women whom they can train, and often develop a complex relationship with their servants based on both dependency and exploitation. Middle-class, professional working women hand over their burden of house-work and childcare to lower-class women but then are dependent on these servants. Thus the two circuits of formal and informal employ-ment are intertwined, especially for women.

Working in private households, maids are unlikely to be protected by employment legislation, may be expected to work very long hours and may also be exposed to the sexual advances of the male members of the household. If they become pregnant they will lose their job and may have to turn to prostitution. Sometimes they will return to their natal rural villages to get married or may stay in the cities and find better-paid jobs.

In Southeast Asia the provision of sexual services employs many women especially in areas with large foreign military bases and a tradition of men seeking sexual gratification outside marriage. Some countries, such as the Philippines and Thailand, have even developed a tourist industry which exploits the trade in female sexuality. Today many countries do not enforce laws against prostitution because of its importance to tourism. Numbers of women involved are difficult to obtain but it is estimated that there are between 100,000 and 200,000 in Bangkok working in 977 establishments in the city. (See *Tourism and Development in the Third World* by John Lea in this series.)

Case study I

Dreams of riches outweigh fear of Aids in Korea

"I only have one body, but I hope to get rich during the Olympics," said a 23-year-old prostitute, Kyong-a. This dream of riches is typical of prostitutes in South Korea where, on some estimates, such women outnumber soldiers.

While a few prostitutes are college graduates, most are from poor families. Many leave home to escape sexual or physical abuse. Others work to help support siblings. A significant number are unmarried mothers whose sense of shame leads them to the profession. Some are abducted by gangsters, raped and forced into prostitution.

Dr Helen Chu, director of a Public Health Centre, holds meetings with the prostitutes. She has declared that she will not obey a new Prevention of Aids Law requiring that Aids victims and carriers be gaoled for three years for unauthorized sexual activity. "I have persuaded many of them to give up. They are now beauticians or housewives," she said. She has also arranged for the authorities to give the women a 50 per cent discount on a taxi licence and many have become drivers.

In male-dominated Korean society, where couples were traditionally forced into arranged marriages by their parents, marital infidelity by husbands is the norm. Most prostitutes work in small inns with a madam and two to five other girls. Old women act as pimps and bring in the customers. The government does little to enforce the law against prostitution and has even been accused of encouraging the trade to earn foreign exchange.

Source: M. Breen (1988) 'Olympics fuel a dream more potent than Aids', *Guardian* 26 July

For poor, uneducated women migrants to the city, prostitution offers the best-paid employment and may be the only job which enables them to send money back to their families in the rural areas. Parents may be forced by poverty to sell young girls to the secret societies who control prostitution in many Asian cities. In South and Southeast Asia, where there were as many as 42 million children under 15 working in 1975, young people are very vulnerable and prostitution may seem easier and more profitable.

In African cities prostitution is less organized and many of the women work on their own account. This may enable women to keep more of their earnings but also means they are offered less protection. Aids is already a greater problem in Africa than in Asia. A recent study in Nairobi, Kenya, showed that the proportion of prostitutes who were HIV-positive rose from 4 to 59 per cent between 1980 and 1986. It remains to be seen what long-term impact this disease has on this female-dominated profession.

The imposition of Western consumption patterns on traditional societies causes conflict. Third World governments, faced with falling prices for commodity exports and a balance of payments crisis, may encourage patriarchal family control of women in order to offer foreign companies and male tourists cheap, skilled young women. Thus national prosperity is seen to depend on a continuation of female subordination and poverty.

Key ideas

1 Although male economic activity rates are spatially homogeneous, rates for women have distinct regional patterns.
2 Women are moving into industry faster than men, especially in Asia.
3 Multinational companies are attracted to Third World countries as sites for export-oriented factories by the availability of cheap, docile, female labour.
4 Female employment in the modern sector may be economically necessary but often conflicts with traditional social values.
5 The constraints of childcare and patriarchal control of women's spatial freedom force women to take on badly paid productive work in the home as out-workers or petty commodity producers.
6 In the service sector women dominate as domestic servants in Latin America, as traders in West Africa and the Caribbean and as prostitutes, most conspicuously in Southeast Asia.

7 Women are often forced to seek employment opportunities in marginal, often illegal occupations such as prostitution or the making of alcoholic beverages, where the financial gain does not always outweigh the risk.

7
Women and development planning

One of the themes running through this book is the failure of development planners to consider women's needs and women's viewpoints.
Women are central to development. They control most of the non-
money economy through bearing and raising children, and providing
much of the labour for household maintenance and subsistence agriculture. Women also make an important contribution to the money
economy by their work in both the formal and informal sectors.

Everywhere in the world women have two jobs – in the home and
outside it. Women's work is generally undervalued and the additional
burden development imposes is usually unrecognized. Their health
suffers, their children suffer and their work suffers. Development itself
is held back. As Third World countries face new problems women's role
becomes increasingly central.

Contemporary problems of development

The environment

It has been said that the present paradox of development lies in the
mistaken assumption that growth of commodity production will improve
the satisfaction of basic needs. Yet the expansion of cash-cropping and
production for export has not been accompanied by the trickle down of
benefits to the poor, especially poor women, while at the same time it
has led to water pollution, soil erosion, destruction of firewood resources

and loss of genetic diversity of plant and animal stocks. A new impoverishment of women has been brought about by the absorption into the market economy of much of the natural resources of land, water and timber on which family subsistence depended, without offering women a new means of support. Their responsibility for maintaining the family by providing food, water and firewood for cooking and heating makes women very aware of environmental degradation and determined to do something about it. To achieve sustainable development, with human and natural resources brought into a dynamic equilibrium, the skills and knowledge of the women, who are the primary sustainers of society, must be utilized.

Case study J

Kenya – women's role in reafforestation

In 1989 Professor Wangari Maathai was awarded Woman-Aid's 'Woman of the World' award. This was in recognition of her work in setting up the Green Belt Movement in Kenya. During 14 years of work with the National Council of Women she persuaded communities throughout Kenya to plant more than 10 million trees. Now 35 other African countries have taken up the scheme.

She started the movement despite opposition from 'experts' who said that all that was needed was greater production of cash crops for export and more fertilizer to produce more intensively. Professor Maathai realized that much of Kenya had already been cleared of trees and bushes and that more cash crops would only accelerate the process of desertification. Over the years the land had become degraded and soil erosion had developed into a serious problem. In most parts of the country there was little firewood left to gather for heating or cooking. The land, quite simply, was unable to support the people. Rural Kenyans were forced to depend on agricultural residues and dung for cooking and heating and to eat an increasing number of highly processed foods. Thousands of rural dwellers flocked to the cities. "The slums are just a symptom", says Prof. Maathai, "of the degradation of the land. The problems of today are totally connected to yesterday's environmental degradation. Economic investment was the god when the problem was really one of simple things like firewood, food and topsoil."

Case study J *(continued)*

Thousands of green belts have now been planted and many hundreds of community tree nurseries set up. Women have shown each other how to collect the seeds of nearby indigenous trees, and how to plant and tend them. Slowly the devastating effects of soil erosion are being reversed.

Source: John Vidal (1989) 'Root causes, root answers' *Guardian* 3 November

Such active resistance to environmental degradation has forced policy makers to rethink the relationship between development and the environment. The women who lead ecology movements in the Third World are offering a new view of development. The Chipko forest conservation movement in India pre-dated the UN Decade for Women. It became identified with women as it was they who led the fight to protect the forest and so preserve sources of fuel and reduce soil erosion. Firewood is the main source of energy in Africa (see *Primary Resources and Energy in the Third World* by John Sousson in this series) but it is becoming more difficult to obtain as land is cleared for cash-cropping. In Ghana, as shown in Figure 7.1, for many families the local supply of wood is now inadequate. Women in households in the savanna have to walk further in search of fuel now than 10 years ago, while over half of those in the forested zone are for the first time experiencing problems, with almost 30 per cent coping with an inadequate supply of firewood. Women have adopted two main coping strategies: reducing the number of cooked meals to one every other day as in parts of West Africa and the Andes, resulting in lower family nutrition levels; and supplementing or replacing fuelwood with agricultural residues such as cassava stalks or dried dung. Some 800 million people now rely on these residues for part of their energy needs thus depriving the soil of inputs traditionally used to improve fertility and soil structure. Many of women's income-earning activities such as fish-smoking, beer-brewing and pottery-making also depend on adequate supplies of fuelwood. In addition, forests supply other raw materials and food products import-ant to women for household consumption, animal fodder and as a source of income.

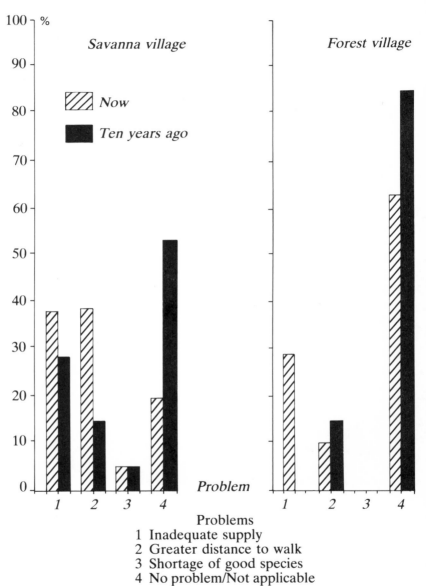

Figure 7.1 Ghana: problems of fuelwood collection in different ecosystems
Source: Ardayfio-Schandorf, 'Rural energy supply and women's work in Ghana', in J. H. Momsen (ed.) *Geographical Studies in Women and Development*, (provisional title) to be published by Routledge

The spread of irrigation, usually associated with green revolution crops, has led to a decline in both quantity and quality of drinking water. In the Indian state of Madhya Pradesh in 1980 out of 70,000 villages, 36,420 reported water shortages; in 1982 this number rose to 50,000 and in 1985 to 64,565. In some parts of Uttar Pradesh state women have to walk 24–32 kilometres to find drinking water. The burden on women is becoming so acute that they are refusing to marry men from water-shortage areas. The Chipko movement is also trying to protect water sources as it is aware of the link between forest and the maintenance of groundwater levels and water purity. It is now recognized that the exclusion of women from the planning of water supply and sanitation schemes is a major cause of their high rate of failure.

Drought in Africa has caused enormous problems for rural women. Even in dry years there are marked local variations in rainfall and so men find it advantageous to have several wives who farm widely scattered plots. This spreads the risk of crop failure and ensures that the husband with many wives will have at least one wife with a successful harvest. It does not reduce the risk for the other wives as traditionally each wife is responsible for feeding her own children. In some parts of the Sudan women are forced to leave their land and migrate to wetter regions where they work as sharecroppers for wealthy landowners and are paid with a percentage of the harvest which can be used to feed their children and their husband.

Economic restructuring

During the last decade many Third World countries have found themselves with an increasing debt burden. Often they have been forced to ask for financial assistance from the International Monetary Fund. In return for this assistance the IMF usually imposes tough financial constraints and demands that the recipient countries restructure their economies. This structural adjustment generally involves an increase in production for export combined with demand-reducing policies such as removal of subsidies on basic foodstuffs, reduction in welfare services, price rises, wage cuts and job losses. It may be argued that the social costs of structural adjustment would have been worse without the IMF intervention but there is little appreciation by international agencies of the gender bias in their impact. In response women have developed new survival strategies. This behaviour has been called 'the invisible adjustment' implying that women make adjustment policies socially possible by increasing their own economic activity, by working harder and by self-abnegation.

Structural adjustment shifts the burden of welfare from the state to individual families and especially women. Wage cuts force more members of the family to seek paid employment and because women are paid less than men it may be easier for them to find employment. The impact of the economic crisis of the 1980s on female economic activity rates is very mixed. It varies not only from country to country but within countries, between economic sectors and urban and rural areas and according to age and educational levels. The most widespread effect was a slowdown in the participation of women in the formal workforce which had been growing since 1960. Women have been particularly affected by the industrial restructuring brought about as a result of the introduction of new less labour-intensive technology and of decreased foreign direct investment in assembly industries in most of the 40 Third World countries operating export processing zones. Many women have joined the informal economy or have been forced to migrate. In the Philippines the number of women seeking jobs overseas increased by 70 per cent between 1982 and 1987. Whether women work often depends on the availability of a daughter to take on domestic chores and childcare. In Malaysia it was found that 31 per cent of girls but only 17 per cent of boys did work in the home. Girls may be taken out of school to replace the mother and so lose their chance to be trained for a better job in the future. For example in Bangladesh, when an IMF-supervised adjustment programme was implemented in 1985, the ratio of girls to boys in primary education fell from 77 girls per 100 boys in 1984 to below 67. Other countries experiencing such changes were Brazil, Chile, Colombia, and Jamaica while decreased female enrolment in secondary schools has been documented for Indonesia, Jamaica, Nigeria, the Philippines and the Sudan.

Increased food prices force poor families to reduce both the quantity and quality of their food intake and women and children are usually most affected. In a slum in Guayaquil, Ecuador, during the economic crisis of 1987–8, 42 per cent of families gave up drinking milk and 79 per cent of children attending clinics suffered from malnutrition. In a sample of 51 countries studied by UNICEF in 1986 all but two had experienced an increase in child malnutrition between 1980 and 1985 and there is some indication that undernourished children were more likely to be female. Increased infant mortality in the period 1982–5 was reported for Brazil, Ghana, Uruguay and the Philippines and in many areas these changes seem to have affected girls more than boys.

Men migrate to seek work elsewhere with the result that the number of both extended and female-headed households increases. Women become more powerful in the face of such acute threats to the survival of the family because of their traditional responsibility for reproduction. Both mothers and daughters work longer hours and time becomes their scarcest resource. Women seek alternative sources of income to compensate for declines in household income and spend longer shopping for and preparing cheaper types of food. Men feel themselves marginalized and the study in Ecuador showed that adult males responded by increasing their alcohol consumption and their level of violence to women while teenage sons turned to dependence on drugs. The poorest families, often headed by women, bear a disproportionate share of the burden of adjustment and the economic crisis has tended to exacerbate pre-existing gender inequalities.

Community management

Awareness of the needs of their communities tends to be greater among women than men, since it is normally women who have to cope with problems of housing and access to services. Consequently, women often take the lead in demanding improvements in urban services. They may also work together to change social attitudes. Groups of women in Bombay, India, march silently, carrying placards, around houses in which dowry deaths have occurred, bringing public shame on the perpetrators. This is more effective than government legislation in reducing the number of these tragedies. Women's groups in India have also lobbied for legal restraints on the abortion of female foetuses. As pressure on women's time increases, their community management role may change.

Women's survival strategies often depend on building up networks of women within the community. The communal kitchens set up in Lima, Peru, help to reduce the time women spend individually cooking for their families and so allow them extra time to earn money. Development agencies often advocate the spread of these grassroots separate women's organizations because they feel that they avoid confrontation with cultural patterns which oppose the mixing of unrelated women and men and prevent a submergence of women's interests and loss of leadership to men. However, these women's groups may provide a focus for the politicization of women's lives around issues of prime importance to their domestic role, such as rising food costs and the disappearance of their children at the hands of repressive regimes. This

Plate 7.1 Brick making in Mysore, India
Source: Janet Townsend, University of Durham

link between the empowerment of women for household welfare and consequent political action has not been analysed by most development workers.

As governments are forced to cut back on public sector spending, the burden of providing basic needs services to poor communities is falling increasingly on women. Women are having to spend precious time negotiating directly with international agencies and non-governmental organizations in order to get free assistance for their children, and training or credit for setting up income-earning enterprises.

Development projects directed at women are often small, scattered and peripheral to the main aims of development. They usually try to promote greater self-sufficiency rather than development in the sense of expansion and qualitative change. Furthermore, the criteria for success are often less stringent than those for projects specifically for men. On the other hand, when general development projects are planned women may find themselves excluded because of restrictive entry conditions. Female-headed households are numerous in the slums of Third World cities but eligibility for new housing is commonly based on the premise that there is a male head of household. Female household heads may not have an income which is large enough or secure enough to qualify for housing. In self-help projects they may not have the time or skills necessary to build a house and if they employ a man to do it they may be cheated. These problems are now widely recognized and can be overcome through training, membership of a women's group or special eligibility conditions tailored to suit the constraints of women's lives.

The success of the women who organized the rebuilding of their homes after the earthquake in Mexico City in 1986 is a good example. They learned how to lobby politicians, how to design apartments which were suitable for their lifestyles and family size and how to prevent contractors cheating them. In doing so they not only rehoused their families but also successfully challenged the patriarchal structure of households, trade unions and political parties. In the face of an environmental disaster, grassroots women's groups were instrumental in reviving their communities.

Development policies for women

Since the 1950s, when development planning first came into the international spotlight, a number of approaches, having different effects on women, have been tried. They may be detailed as follows.

Welfare

This was the earliest approach. It dominated from 1950 to 1970 and is still widely used. Its main purpose was to enable women to be better mothers which was seen as their main role in society.

Equity

This was the original approach of Women in Development (WID) and was utilized during the Decade for Women 1975–85. Women were seen as active participants in the development process.

Anti-poverty

The second WID approach, it aimed to increase the productivity of poor women and saw their poverty as a problem of underdevelopment not of subordination.

Efficiency

This is the most prevalent approach used today. Its aim is to ensure that development is efficient and effective. It is based on an awareness that policies of economic stabilization and adjustment rely on women's contribution to development and aims to make them more efficient managers of poverty.

Empowerment

An approach articulated by Third World feminists since the mid-1980s. It aims to empower women through greater self-reliance and sees women's oppression as stemming not only from male patriarchal attitudes but also from colonial and neo-colonial oppression.

Efforts to graft an awareness of women's needs onto planning through such political action as the Percy Amendment in the United States, which has been followed by similar policies in the Canadian and British government aid agencies and in the World Bank, have had little effect. It is necessary to challenge planning stereotypes relating to the structure of families and the division of labour within low-income households. Planners are often unaware of the work burdens of women and of their problems of time as a scarce resource. The kind of integrated project women need, such as making health and training facilities available at a convenient time and place for women with small children and little access to transport, is rarely attempted.

It is also necessary to distinguish between the practical and strategic needs of women. Practical needs are those such as food and shelter which are required by all the family and are identified as priorities by women and planners alike. They serve to preserve and reinforce the gender division of labour. Strategic needs are those which can empower women, challenge the existing gender division of labour and bring about greater equality. They are difficult to meet because poor women often do not have time to reflect on such needs because of their immediate requirement of seeking the satisfaction of their practical needs. Because equity programmes disturb the status quo and demand long-term commitment by governments they are rarely implemented. However, addressing women's strategic needs is vital if fundamental change is to occur.

The costs of ignoring the needs of women are many: uncontrolled population growth, high infant and child mortality, a weakened economy, ineffective agriculture, a deteriorating environment, a

Plate 7.2 Aborigine painter. The Australian government at first ignored women's land rights in Aborigine areas. The women formed protest groups and forced planners to reconsider gender differences in the traditional importance of specific sites. Market demand for aboriginal paintings has created new income-earning opportunities for women in this traditionally male-dominated activity
Source: Elspeth Young of the Australian Defence Force Academy, Campbell ACT

divided society and a poorer life for all. For young girls and for women it means unequal opportunities, a higher level of risk and a life determined by fate and the decisions of others rather than by choice.

Conclusion

Economic crisis in many Third World countries, enhanced by their peripheral position in the world economy, has led to reductions in spending on health, education and food subsidies and the impact is heaviest on poor women. When women are able to respond successfully to crises they gain status within the household, either because they have become the chief income earner in the family or because they have gained confidence through learning how to negotiate successfully with national and international agencies, and to work with other women. This very success may provoke an additional crisis in the internal gender relations of the household. Women's increased power and independence may result in a male backlash of violence and the expansion of female-headed households. It may also lead to more equality and freedom of choice for both men and women. The conflict between 'machismo', or male dominance, and economic need is creating societies in a state of flux in many parts of the Third World.

Trying to develop without acknowledging the people who do two-thirds of the work is inviting failure. Ways must be found to reduce their burden of work if women are to realize their potential. Development plans for women, where they exist, tend to assume mistakenly that women have free time to devote to new projects and to ignore the heterogeneity and differentiation of women.

Women are agents of change, not just victims. The United Nations has realized that the role and status of women are central to changes in population and development. It now argues that development plans must be rethought from the start so that women's abilities, rights and needs are taken into account at every stage. Making investment in women a development priority will require a major change in attitudes to development, not only by governments but also by lending agencies.

Investing in women is not a global panacea. It will not put an end to poverty but it will make a critical contribution. And it will help to create the basis for future generations to make better use of both resources and opportunities.

Key ideas

1 Women are central to environmental protection and the success of sustainable development.
2 Women's networks and groups can play an important role in grass-roots projects for community improvement and social change.
3 It is necessary to distinguish between the practical and strategic needs of women.
4 Overwork and shortage of time are often ignored as barriers to women's participation in development projects.
5 Development policy towards women has changed from one which saw them as only mothers to a more holistic approach which emphasizes both the productive and reproductive roles of women.
6 The increased opportunities for women to be economically independent are leading to changes in gender relations.

Further reading and review questions

Chapter 1

1 Why do you think that both international and national policy directives aimed at improving the status of women have had so little effect?
2 Identify and locate some of the countries included in the low-income group and in the high-income oil-exporting group shown in Figure 1.1. How do the figures for the income group compare with the figures for the regional group to which the countries you have identified belong? Why are there both differences and similarities?

Further reading

Boserup, E. (1989) *Women's Role in Economic Development*, London: Earthscan Publications Ltd.

Brydon, L. and Chant, S. (1989) *Women in the Third World. Gender Issues in Rural and Urban Areas*, Aldershot: Edward Elgar.

Momsen, J. H. and Townsend, J. (eds.) (1987) *Geography and Gender in the Third World*, London: Hutchinson.

Radcliffe, S. with J. Townsend, (1988) *Gender in the Third World. A Geographical Bibliography of Recent Work*, Sussex: Institute of Development Studies. Development Bibliography Series No.2.

Seager, J. and Olsen, A. (1986) *Women in the World: An International Atlas*, London: Pan Books Ltd.

Sivard, Ruth L. (1985) *Women: A world survey*, Washington D.C.: World Priorities.

Townsend, J. (1988) *Women in Developing Countries. A Select Annotated*

Bibliography for Development Organisations, Sussex: Institute of Development Studies. Development Bibliography Series No.1.

United Nations Office at Vienna, Centre for Social Development and Humanitarian Affairs (1989) *1989 World Survey on the Role of Women in Development*, New York: United Nations.

Chapter 2

1 Both Montserrat in the West Indies and Lesotho in southern Africa have experienced high male out-migration. Compare the position of the women left behind in these countries and consider the reasons for the differences you have noted.

2 How have marriage arrangements, farming systems and economic development influenced the regional pattern of the sex ratio in India?

3 Females are generally less well-nourished than males in poor Indian families despite the heavy workload of women and girls. Why does this occur and what are the long-term effects on the population?

Further reading

Anker, R., Buvinic, M. and Youssef, N. (eds.) (1982) *Women's Roles and Population Trends in the Third World*, Geneva: International Labour Office.

Hamilton, S., Popkin, B. M. and Spicer, D. (eds.) (1984) *Women and Nutrition in Third World Countries*, London: Greenwood Press.

Morokvasic, Mirjana (1984) 'Women in migration' Special Issue of *International Migration Review*, Vol.18, No. 4, Winter. New York: Center for Migration Studies of New York Inc.

Mortimer, Delores M., and Bryce-Laporte, Roy S. (1981) *Female Immigrants to the United States: Caribbean, Latin American and African Experiences*, Washington D.C.: Research Institute on Immigration and Ethnic Studies, Smithsonian Institution.

World Bank (1989) *World Development Report, 1989*, Oxford: Oxford University Press.

Zeidenstein, S. (ed.) (1979) 'Learning about rural women' Special Issue of *Studies in Family Planning*, Vol. 10, (11/12)

Chapter 3

1 Distinguish between the four aspects of reproduction.

2 Consider the relationship of urbanization, education of women, and rising living standards to changes in fertility rates.

3 How does inadequate housing increase women's workload? How can women in shanty towns improve their living conditions?

Further reading

Black, Naomi and Cottrell, Ann Baker (eds.) (1981) *Women and World Change. Equity Issues in Development*, London and Beverly Hills, CA: Sage Publications Inc.

Catasus, S., Farnos, A., Gonzalez, F., Grove, R., Hernanadez, R. and Morejon, B. (1988) *Cuban Women: Changing Roles and Population Trends*, Geneva International Labour Office. Women,Work and Development 17.

Moser, C. O. N. and Peake, L. (eds.) (1987) *Women, Human Settlements and Housing*, London: Tavistock.

Sadik, N. (1989) 'Investing in women: the focus for the nineties', *The State of the World Population 1989*, New York: United Nations Population Fund.

Shah, Madhuri (ed.) (1986) *Without Women, No Development. Selected Case Studies from Asia of Non-Formal Education for Women*, London: Commonwealth Secretariat Publications.

Chapter 4

1 Why are women so important in African farming?
2 With reference to Figure 4.2 compare the effects on women of structural, technical and institutional changes in agriculture.
3 Redraw Figure 4.3 assuming that the family did not include an adult daughter and that there was a pre-school child at home. Who would take over the marketing and the care of the small child? How would these changes in family structure affect the mother's workload? What jobs could no longer be undertaken? What would be the effect of the following additions to the family: an adult niece or an adult nephew or the grandmother? What does this analysis tell you about the importance of family life stage and dependency ratios (the proportion of working age members to children and elderly people) and the labour productivity of the family?

Further reading

Ahmed, I. (ed.) (1985) *Technology and Rural Women*, Geneva: International Labour Office.

Beneria, L. (ed.) (1982) *Women and Development. The Sexual Division of Labour in Rural Societies*, New York: Praeger Special Studies.

Deere, Carmen Diana (1982) 'The division of labour by sex in agriculture: a Peruvian case study', *Economic Development and Cultural Change*, Vol. 30 (4). pp. 795–811.

Dixon-Mueller, R. (1985) *Women's Work in Third World Agriculture: Concepts and Indicators,* Geneva: International Labour Office. Women, Work and Development 9.

Chapter 5

1 To what extent is it true to say that equality of education is a necessary but not a sufficient condition for equality of pay?

2 The Caribbean data in tables A and B illustrate the problems of learning about women's work from national censuses. Table A lists categories of work identified by a sample of adult women in villages in the Leeward Islands of the Caribbean. Table B provides the occupational categories used in the 1980–1 Population Census of the Commonwealth Caribbean.

Table A	Table B
Village women's work	*Occupational classification*
Leeward Islands	*Caribbean census*
housework	professional/technical workers
goldsmith/shopkeeper	administrative/managerial
hotel maid/farmer	clerical and related work
bank clerk	sales workers
plantation labourer	service workers
cook/farmer	agricultural and related workers
office worker	production and related workers
maid/part-time inter-island trader	not stated
labourer on government farm/ housework	
potter/housework	
domestic servant/farmer	
shopkeeper/farmer	
baker/farmer	
domestic servant/housework	
garment factory worker	
teacher	
sells vegetables in market on neighbouring island/farmer	
shop assistant	
housework/dressmaker	
hotel receptionist	
works in airline office	
cook at airport	
sells cooked food from home	
sells vegetables in local market	
picks cotton/housework	
civil servant/farmer	

a) Try to allocate all the women's work listed in Table A to the categories used in the census. Each person may be allocated to only one category.
b) What kinds of work are easy to allocate?
c) What kinds of work present problems?
d) How would you modify the census classification to give a better description of women's work?

Further reading

Anker, R. and Hein, C. (1986) *Sex Inequalities in Urban Employment in the Third World*, London: Macmillan.

Dixon-Mueller, R. and Anker, R. (1988) *Assessing Women's Economic Contributions to Development*, Geneva: International Labour Office. World Employment Program Paper No. 6.

Joekes, Susan P. (1987) *Women in the World Economy*, an INSTRAW Study. Oxford: Oxford University Press.

Young, Kate and Moser, Caroline O. N. (1981) *Women and the Informal Sector*, Institute of Development Studies Bulletin, 12 (13). University of Sussex.

Chapter 6

1 From a study of Figures 6.1 and 6.2 identify two regions with high levels of female employment. Consider the influence of political system, type of agriculture and industrial base on the employment of women.
2 Outline the economic, social and cultural reasons for the recent expansion of prostitution in the Third World.
3 Why is a pool of female labour attractive to world market factories?

Further reading

Beneria, L. and Roldan, M. (1987) *The Crossroads of Class and Gender: Industrial Homeworkers, Subcontracting and Household Dynamics in Mexico City*, Chicago and London: University of Chicago Press.

Heyzer, Noeleen (1986) *Working Women in South-East Asia*, Milton Keynes: Open University Press.

Leacock, E. and Safa, Helen. I. (eds.) (1986) *Women's Work. Development and the Division of Labour by Gender*, Massachusetts: Bergin and Garvey Publishers Inc.

Nash, June C. and Fernandez-Kelly, Maria Patrica (eds.) (1983) *Women, Men and the International Division of Labour,* Albany: State University of New York Press.

Chapter 7

1 'Women are victims of environmental crisis.' Discuss.
2 What is meant by household survival strategies? Give examples.
3 You are asked to help implement three development projects:
   a) A new health care programme in Nepal.
   b) A self-help housing project in Jamaica.
   c) The building of a road between an isolated village and the capital city in Bolivia.
   Consider the following points:
   i) Would you identify a 'special aspect' for women in each of the projects?
   ii) How would the involvement of local women and men in a large capital project, such as a road, differ from that of a 'self-help' project?
   iii) What steps would you take to ensure that both sexes had access to the benefits of the projects?
   iv) Suggest ways in which these projects might either improve or worsen women's status, standard of living and workload.
   v) Would these projects meet women's strategic needs or their practical needs?

Further reading

Agarwal, Bina (1986) *Cold Earth and Barren Slopes: Woodfuel Crisis in the Third World*, California: Riverdale.

Dankelman, Irene and Davidson, Joan (1988) *Women and Environment in the Third World*, London: Earthscan Publications Ltd.

Moser, Caroline O. N. (1989) 'Gender planning in the Third World: meeting practical and strategic gender needs', *World Development.* 17 (11).

Palmer, Ingrid (1985) *Women's Roles and Gender Differences in Development: Cases for Planners*, West Hartford, Connecticut: Kumarian Press.

Rogers, Barbara (1980) *The Domestication of Women: Discrimination in Developing Countries*, London: Tavistock.

Sen, Gita and Grown, Caren (1988) *Development, Crises and Alternative Visions*, London: Earthscan Publications Ltd.

Tinker, Irene (ed.) (1990) *Persistent Inequalities: Women and World Development*, New York and Oxford: Oxford University Press.

Index